NEUROSUTRA

Abhijit Naskar is a celebrated neuroscientist, who became a beloved author all over the world with his first book The Art of Neuroscience in Everything. The book hit the bestsellers list within a few months of publication and heralded the advent of a rejuvenating scientific philosophy of the human mind. The purpose of this philosophy was to enrich human life with scientific sweetness. This book is a collection of his first five books that represent the incredible scientific philosophy of self-awareness.

'Naskar is a self-trained scientist and thinker who discovers the paradigm-shifting phenomena of the human mind in this book.'

Michael A. Persinger, Director of Laurentian University's Consciousness Research Laboratory

'Although this book is full of very useful knowledge and advices, it is also very accessible and fundamental. Understanding how the brain works is also understanding who you are and what your particular skills are. This is a fantastic summary of the amazing discoveries on the brain and a guide to apply them in your personal life by a talented author and a brilliant neuroscientist.'

Ronald Cicurel, co-author of 'The Relativistic Brain: How it works and why it cannot be simulated by a Turing machine'

'The book is very interesting and useful. I am sure Neurosutra will be very timely and effective.'

Sam Pitroda, Father of Indian Telecom Revolution

THE ABHIJIT NASKAR COLLECTION

NEURO
SUTRA

An Amazon Publishing Company, 1st Edition, 2015

Printed in United States of America

ISBN-13: 978-1516804535

CONTENTS

Preface

Human Brain is the most complex organized structure in the universe. It is a biological organ with mind-boggling capabilities. It is made up of a hundred billion nerve cells or neurons. The relentless activity of these neurons gives rise to all the richness of our conscious experience. All our feelings, emotions, thoughts, ambitions, love lives, religious sentiments and even our sense of self are simply born from the activity of these little specks of jelly in our head. And the field of Science that studies the Brain is called Neuroscience.

The tree of Neuroscience has two branches. One is neurology and the other is psychology. Neurology is the hardware part of the brain, whereas psychology is the software. When these two come together, the magic happens. And this magic is perhaps the most amazing thing on planet earth. It is the magic of a blossoming human life.

No computer can run without the presence of both hardware and software. Likewise, the most complex and fascinating living computer on this planet, i.e. the human brain requires the beautiful and orderly interplay of its neurological circuits and their functional product, i.e. the mind.

The human mind is the most fantabulous creation of the brain circuits. It even allows a person to rewire various regions of the brain based on the daily needs. As a whole, despite being the creation of the human brain, the mind runs the human biology throughout the entire lifetime.

Quite surprisingly the human biology drives the mind as well. It's a kind of surreal twosome between the biology and the mind that keeps a person on his or her feet since birth till death. And once, the underlying crafty mechanism of the human mind is understood properly, life becomes much more cheerful, meaningful and colorful. The unfathomable universe of the human mind holds the key to a better living.

* * *

BOOK I

THE ART OF NEUROSCIENCE IN EVERYTHING

NEUROSUTRA

Introduction

Mankind is the most mysterious of all species on our beloved planet earth. Its every behavior, every action and every single phrase it utters, undeniably show the excellence, with which Mother Earth has molded the human species throughout the entire evolutionary period.

Modern man has come a long way, since he first learnt to control and create fire. He has advanced in various fields like science, arts and technology. Even so, it's the privilege of only a neuroscientist to observe and explore the biological or more specifically neurological functions behind all those advancements. My work is to find out what makes the humans so special and unique among all the species on this planet.

Humans are the only species on earth who can contemplate the vastness of the cosmos, the beauty of the sea, peacefulness of the full moon, the craftsmanship of Mother Nature and even contemplate itself contemplating. We can observe the craftsmanship of Mother Nature most significantly in the evolution of human brain. This 3 lbs. lump of jelly is the driving force, the alpha and the omega of all human excellence.

With this tiny piece of biological instrument we are even able to see galaxies that are light-years away from our dearest Milky Way. Alongside such scientific excellence, most of humanity has also accepted the existence of a Supreme Being, which is usually referred to as "GOD", based on faith and historical records of encounters. This gives us an amazing evolutionary trait or behavioral characteristic called "Faith" to explore.

All experiences, behaviors, beliefs, feelings such as faith, love, attraction, lust, hatred, excitement, kindness, empathy, good and evil that make us humans, are the creation of various intricate and inexplicable molecular interaction within the brain. This book is about those interactions that impact over daily human activity and behaviors. In this book, I'll open up to the reader, the beautiful maze of the brain that creates human experiences, feelings and beliefs in the most non-technical way possible. This book will elaborate on the biological or rather neurological foundation of human mind's deepest instincts, emotions and mysteries, and make it coherently understandable even to the layman.

*　*　*

CHAPTER 1
YOU DRIVE ME CRAZY LIKE HELL - THE NEUROBIOLOGY OF LUST, ATTRACTION, LOVE AND ATTACHMENT

Love is patient and kind. Love is not jealous or boastful or proud or rude. It does not demand its own way. It is not irritable, and it keeps no record of being wronged. It does not rejoice about injustice but rejoices whenever the truth wins out. Love never gives up, never loses faith, is always hopeful, and endures through every circumstance ... love will last forever!

Corinthians 13:4-8, Holy Bible

These few lines from the Holy Bible are so beautiful and perfect. As if all the explanation you need for your overwhelming feeling towards the loved one is there. Love is really an amazing thing. Perhaps it is the most amazing essence of being a human. All animals do it. All of you definitely have fallen in love at least once in your lifetime. The first time, you laid eyes on your dearly beloved, you suddenly started to feel the soothing breeze brushing against your skin, hear the sweet chirping of the birds, even the time seemed to have ceased for you two. For those of you who have recently fallen in love, you'd know even more clearly what I'm saying. As they say, there's nothing crazier than love. And it's really worth being crazy in love. Evolutionary speaking it works as a great motivator, as when you are in love more Oxygen rushes to the brain. And more Oxygen in the brain means more cerebral activity. But there's more to it than just increased O2 supply.

Love is a really complex neurobiological phenomenon, involving lust, attraction, trust, belief and pleasure activities within the brain. But let's keep it as much non-technical as possible and explore the chemistry of love as interestingly as making love.

What is Love?

Do we really want to know! As long as it keeps us up on our feet, who cares!!!

Every human being has his or her own perspective about love. A thousand people would define love in a thousand ways. Attachment, commitment, intimacy, passion, possessiveness, grief upon separation, and jealousy are but a few of the emotionally loaded terms used to describe the representation of love. In a lover's words: *"love is as vast as the cosmos, it is endless, limitless and unconditional"*. Although, if we start exploring the neurobiological mechanism of love with science, it becomes even more exciting and multi-dimensional. If you observe closely enough the astounding neurobiology of love itself can make your brain filled with O2.

This beautiful neurobiological mechanism critically involves various neurochemicals like oxytocin, vasopressin, dopamine, serotonin and endorphins. These neurochemicals make you love and feel loved.

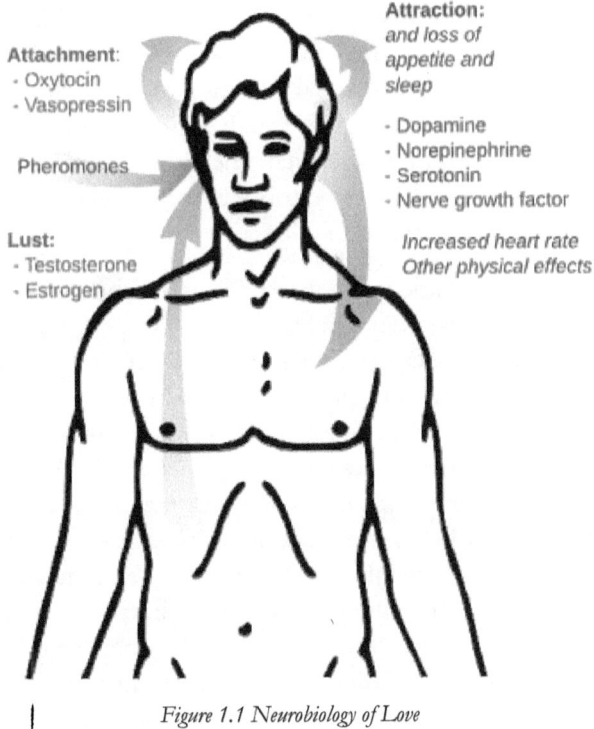

Attachment:
- Oxytocin
- Vasopressin

Pheromones

Lust:
- Testosterone
- Estrogen

Attraction:
and loss of appetite and sleep

- Dopamine
- Norepinephrine
- Serotonin
- Nerve growth factor

Increased heart rate
Other physical effects

Figure 1.1 Neurobiology of Love

The concept of love involves having an emotional bond to someone for whom one yearns, as well as having sensory stimulation that one desires. The word "love," however, derives etymologically from words meaning "desire," "yearning" and "satisfaction" and shares a common root with "libido" (sex drive).

Thus, the psychological sense of love can be interpreted as referring to the satisfaction of a yearning, which may be associated with the obtaining of certain sensory stimulation. Love therefore possesses a close connection not only with reward and pleasure phenomena, but also with appetitive and addictive behaviors. This is the usual addiction of love, when you feel so much addicted to the person you love. Separation from the beloved one leads to withdrawal like symptoms, because your body literally keeps craving for the pleasure stimulation of being in love.

Love and its various emotional states and behaviors are rarely investigated by scientific means. In part, this may be due to the fact that love has always been the domain of poets, artists and philosophers. Although, for ages love has inspired many scientists in their extraordinary discoveries, it has certainly not been considered to be right within the scope of common experimental science, i.e., neurobiology research. Emotions and feelings such as attachment, couple and parental bonding, and love, neglected for centuries by the experimental sciences, have now come into the focus of neuroscientific investigation in order to elucidate the biological mechanisms and pathways of the most pleasing concept of humankind. So let's explore the ingredients of the love phenomenon.

Love begins with the stage of primitive lust and attraction. I'm saying primitive because at this very early stage there is really no difference between primitive man and modern man. The bodily characteristics of a person such as, how hot they are, poke the level of sex hormones (testosterone and estrogen), cortisol and pheromones. Lust is initiated at this stage through the physical attraction and flirting. This is an evolutionary behavior of mankind that biologically enables a human to find a healthy, fertile and perfect mate.

Following the cue of lust, the major attraction symptoms kick in, which are usually known as the symptoms of love, such as sweaty palms, tremors in the whole body, restlessness, loss of appetite and sleep, thumping heart, butterflies in the stomach etc. Such symptoms occur because the body is flooded with neurochemicals like Dopamine, Cortisol, Norepinephrine, and Phenylethylamine (PEA). Once this euphoria wears off, the ultimate and deepest stage of love prevails that is the attachment phenomenon. And the chemicals that make this possible are Oxytocin, Vasopressin and Endorphins. As time goes by, the crazy love sensation diminishes and the feeling of closeness and attachment grows and prevails till the last breath of life.

Love is more than just a blissful Kiss. It is a really pleasurable experience, moreover it's the best natural motivator. Love turns on the reward center of your brain like a light bulb. Everything around you seems so beautiful. You feel an amazing bliss with your beloved partner. As a result deeply intriguing expressions of love come out of your mouth: *"You and I are forever", "You drive me crazy like hell", "I belong to you".*

The intensely sensational and emotional state of love has inspired artists to create masterpieces and even scientists to come up with world-changing ideas like Erwin Schrodinger's Wave Equation and Stephen Hawking's Hawking Radiation. That's why we neuroscientists are so fascinated with the love phenomenon.

Figure 1.2 Obverse side of the old Austrian 1000 Schilling note with the Austrian physicist Erwin Schrodinger

The exhilarating sensation of love cannot be tamed by any law of the sophisticated human civilization. When it overwhelms a person with amorous desires it does not care about whether the person is a philosopher, physicist or a neuroscientist. Take the Austrian Physicist Erwin Schrodinger for example. He is a legend of Quantum Mechanics. Responsible for major advancements in quantum mechanics, Schrodinger's name is now synonymous with what is known as the "Schrodinger's cat" paradox, a thought experiment in which a cat, trapped in a box with a breakable vial of poison, can be considered both alive and dead by the outside observer until the box is opened, illustrating the principle of superposition in quantum theory.

In 1926 Schrodinger presented a theory explaining the spectrum of the hydrogen atom, likening its functioning to a wave equation, thus laying the foundation for wave mechanics. His ideas of wave mechanics were inspired by de Broglie's ideas, which he had first thought to be completely rubbish, until he was persuaded otherwise. Now comes the juicy part, in the words of the physicist Hermann Weyl, Schrodinger obtained his inspiration for wave mechanics while engaged in a *"late erotic outburst in life"*. He was what we can call the greatest Casanova among all scientists. To be more specific, he had a wandering eye for young girls on the brink of adolescence. In the book Erwin Schrodinger and the Quantum Revolution, biographer John Gribbin portrays the man as: *"He was often in love – or he convinced himself that he was in love – and when he was in love, by and large life was good and his scientific creativity benefited."*

In 1920 he married Annemarie Berthel and despite all the stormy episodes in the relationship they remained together until his death. He had no children by his wife, but he had at least three illegitimate daughters.

One wonders whether Schrodinger's famous wave equation was a product of marital bliss. It was indeed a product of love, but there's a twist in the story. As I said earlier he was an amorous philanderer. His wave equation, the key equation of quantum physics emerged during one of his amorous adventures in late 1925, when he stayed at a hotel resort with a mistress while his wife remained in Zurich. The woman involved in this particular encounter has not been identified, but whoever she was, she deserved to be admired as the *"greatest muse of a great thinker"*.

Figure 1.3 Tombstone of Annemarie and Erwin Schrödinger (1887–1961) with the Schrodinger equation

Arnold Sommerfeld once referred to the Schrodinger's wave equation as: *"the most remarkable of all remarkable discoveries of the 20th century"*. When Max Planck held Schrodinger's second publication in his hands, he sent Schrodinger a post card saying:

"I am reading your communication in the way a curious child eagerly listens to the solution of a riddle with which he has struggled for a long time, and I rejoice over the beauties that my eye discovers, which I must study in much greater detail, however, in order to grasp them entirely."

With reference to the origin of the Schrodinger equation, the American Nobel laureate Richard Feynman noted:

"Where did we get that [Schrodinger's equation] from? Nowhere. It is not possible to derive from anything you know. It came out of the mind of Schrodinger, invented in his struggle to find an understanding of the experimental observation of the real world."

So, the bottom-line is that the erotic adventure of Erwin Schrodinger led to the birth of the key equation of quantum physics. He won the Nobel Prize for Physics in 1933 for his 1926 introduction of the wave equation. He was a true genius of modern physics, but his innate biological urges didn't know the bounds of the society. He could not keep his hands of teenage girls. He had children by at least three of his mistresses, including a daughter by Hilde March, the wife of his colleague Arthur March. His wife Anny once said: *"it would be easier to live with a canary bird than with a racehorse, but I prefer the racehorse."* In the face of primitive urges, even the great physicist turned out to be helpless. That's how strong the early stage of lustful attraction can be.

Molecular signaling can make the experience of early love pleasant and stressful at the same time. Until you really express your feelings to the person, there remains a sort of uncertainty, which is kind of stressful. This uncertainty leads to a duel inside your mind and therefore increases the cortisol level. Once you spill the eggs, the stress hormone cortisol level goes down.

The very first phase of love gives rise to a crazy euphoric sensation. The passion of love creates the feeling of exhilarating happiness that is often unbearable and certainly indescribable. And the areas that are activated in response to romantic feelings are largely co-extensive with those brain regions that contain high concentrations of a neuro-modulator called dopamine that is associated with reward, desire, addiction and euphoric states. Like two other modulators, oxytocin and vasopressin that are linked to romantic love, dopamine is released by the hypothalamus, a structure located deep in the brain and functioning as a link between the nervous and endocrine systems.

These same regions become active when exogenous opioid drugs such as cocaine, which themselves induce states of euphoria, are ingested. Release of dopamine puts one in a "feel good" state, and dopamine seems to be intimately linked not only to the formation of relationships but also to sex, which consequently comes to be regarded as a rewarding exercise. An increase in dopamine is coupled to a decrease in another neuro-modulator, serotonin (5-HT or 5-hydroxytryptamine), which is linked to appetite and mood.

Love, after all, is a kind of obsession and in its early stage commonly immobilizes thought and channels it in the direction of a single individual. At this early stage you cannot think of anything else but that apparently perfect special person. Different regions of the brain mostly analytical and logical ones like the prefrontal cortex become less active. That explains why people who are in love are not able to judge their partner's character honestly. Come on, let's be honest, we have all felt like this, when we first fell in love – as if we look at the special person "Through Rose-Colored Glasses".

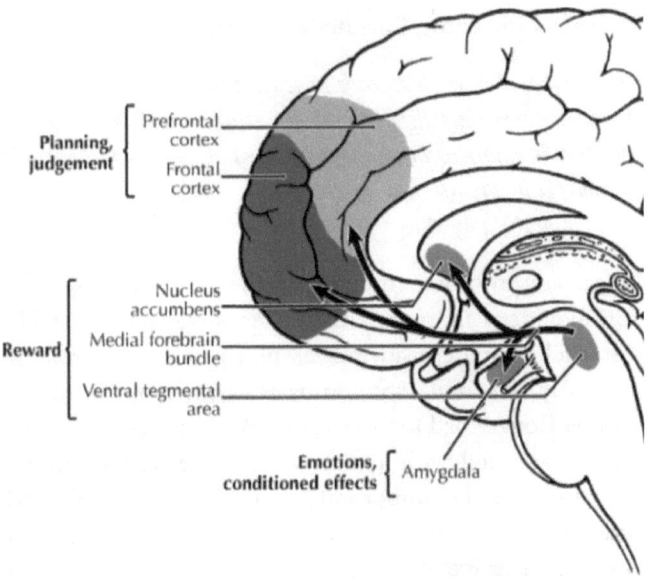

Figure 1.4 Brain Areas involved in pleasure/reward, judgment and emotional effects

Euphoria and suspension of judgment can lead to states that others might interpret as madness. It is this madness that poets and artists have celebrated.

Plato considered it in "Phaedrus" as a productive, desirable state. But this cognitive blindness towards your beloved one's dark side at the first stage of romance has an evolutionary importance. It helps you pass the mad love stage without being confused by any behavioral response from the other. At the beginning of a romantic relationship both the individuals pose to be anything but themselves. It's not until the later attachment and bonding stage when the two individuals start to open up to each other.

In Plato's Phaedrus, Socrates comments:

"the irrational desire that leads us toward the enjoyment of beauty and overpowers the judgment that directs us toward what is right, and that is victorious in leading us toward physical beauty when it is powerfully strengthened by the desires related to it, takes its name from this very strength and is called love".

There are no moral strictures in the early stage of love, for judgment in moral matters is suspended as well. After all, moral considerations play a secondary role, if they play one at all, with Anna Karenina, or Phedre, or Emma Bovary or Don Giovanni. And morality, too, has been associated with activity of the pre-frontal cortex. So it is really important that the analytical and judgmental pre-frontal cortex is partially switched off, in order to survive the early stage of intense and crazy sensation of lust, crush, infatuation, attraction and obsessiveness. The early stage of romantic love seems to correlate as well with another substance, nerve growth factor, which has been found to be elevated in those who have recently fallen in love compared to those who are not in love or who have stable, long-lasting, relationships. Moreover, the concentration of nerve growth factor appears to correlate significantly with the intensity of romantic feelings.

As time passes, the euphoria of love wears off and serotonin level gets back to normal. Now begins the most mature stage of true love, that is attachment and bonding through closeness and sexual intimacy. Here passionate love transforms into compassionate warm love. The ingredients of this stage are Oxytocin and Vasopressin. Alongside the feel good chemical Dopamine, these two neuro-modulators are the most prominent hormones in attachment and bonding. Both are produced by the hypothalamus and released and stored in the pituitary gland, to be discharged into the blood, especially during orgasm in both sexes and during child-birth and breast-feeding in females. In males, vasopressin has significant link with social behavior, in particular to aggression towards other males. The concentration of both neurochemicals increases during the phase of intense romantic attachment and pairing. The receptors for both are distributed in many parts of the brain stem which are activated during both romantic and parental love.

Parental love is another crucial ecstatic experience of the human species. Oxytocin, Vasopressin and Prolactin are the major chemicals of parental love and bonding. Oxytocin is one of nature's chief molecules for molding a mother. Roused by the high level of estrogen during pregnancy, the number of oxytocin receptors in the expecting mother's brain multiplies dramatically near the end of her pregnancy. This makes the new mother highly responsive to the presence of oxytocin. It is the key component to ecstatic and orgasmic childbirth. Oxytocin's first important surge occurs during labor. If a cesarean birth is necessary, allowing labor to occur first provides some of this bonding hormone surge and helps ensure a final burst of antibodies for the baby through the placenta. Pressure against the birth canal further heightens oxytocin levels in both mother and baby.

Vasopressin induces the protective nature in a father for his partner and child. During pregnancy testosterone level starts to drop in the father to be. As a result, it reduces the aggressiveness in the male. And prolactin level increases in the father which brings the paternal caregiving instincts. Elevated prolactin levels in both the nursing mother and the involved father cause some reduction in their testosterone levels, which in turn reduces their libidos. Therefore, thanks to Mother Nature's amazing craftsmanship, parental love becomes an ecstatic experience just like romantic love.

All that two lovers want, is to be one with the other in a moment of ecstasy. They want to be ONE in every way possible, being entangled in each other. Sexual union is the closest that humans can get towards achieving that unity. Through sexual intimacy, two human beings become one mind, body and soul wrapped up in the cocoon of their skin.

Love can literally transform a human being. It can make us do either heroic or evil deeds. It is the best inspiration ever. Artists have created various masterpieces inspired by love. William Shakespeare was one of those inspired ones. In "A Midsummer Night's Dream" he expresses romantic love as a dynamic power that always finds a way to overcome obstacles, is not straight-lined and alternates between phases of high and low current.

When you are in love (not in the early crazy love stage), you really are able to achieve greater things, due to increased cerebral activity and the feel good stimulant. While on the contrary, a break up leads to reduced brain activity. If a relationship comes to an end, it is usually experienced as an unpleasant event, with increased levels of stress hormones. Recent studies of brain activity patterns have shown increased activity in areas active during choices for uncertain rewards and delayed responses, reflecting a common feeling of uncertainty about the future. Rejected individuals showed a decreased activity in brain networks involved in the onset of major depression and also showed depressive symptoms, suggesting that the grieving period following a break up might be a major risk factor for clinical depression.

To understand the components of "perfect love" and the "perfect couple", Robert J. Steinberg of the Yale University postulated a theory called the "Triangular Theory of Love". This theory holds that love can be understood in terms of three components that together can be viewed as forming the vertices of a triangle.

These three components are intimacy, passion and commitment.

The complete form of love or "consummate love" consists of all the three components in a balanced ratio, giving rise to the "perfect couple" in a long-term relationship. Such couple will continue to have fifteen or more years of great sex. They cannot imagine themselves happier over the long-term with anyone else. They overcome their difficulties gracefully, and each delights in the relationship with the other. A state that sounds desirable. Maintaining this state is highly dependent on a successful translation of the components into action.

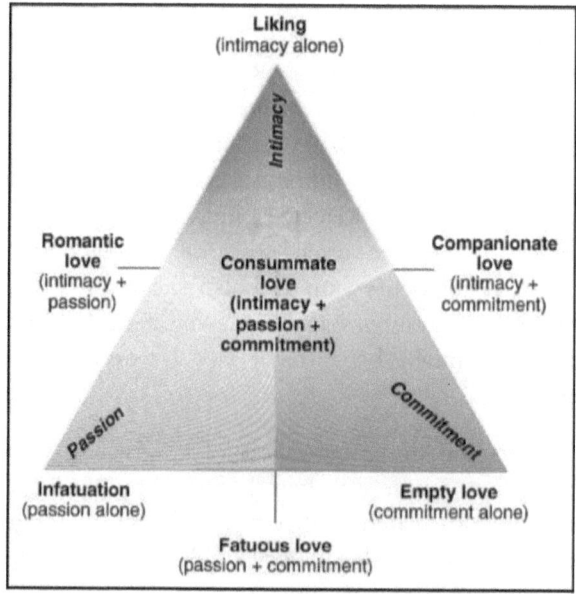

Figure 1.5 The Triangular Theory of Love

Consummate love may not be permanent. And remember not to cross the limit of your body's stress level while having blissful sex, because after that it'd become less pleasuring and more stressful, as then the release of dopamine reduces and the level of stress hormone cortisol rises.

A fulfilling long-term relationship is not accomplished by just finding "the one". It is rather a co-operation between two passionate and highly motivated partners working together, figuring out every single situation holding hands. If there is trust at the root of the relationship, if the partners make an effort to keep it interesting, if difficulties are handled tactfully and if you can appreciate every single deed of your partner no matter how insignificant it is, the flames of love would never burn out and your love can truly "live happily ever after".

Special. Note:

A special tip for those in love. Chocolate is a wonderful romantic gift for your loved one as it contains Phenylethylamine (PEA), the love molecule which induces euphoria and pleasure in the brain.

* * *

CHAPTER 2
CONNECTIVITY OF MINDS - THINK OF THE FRIEND AND THE FRIEND APPEARS

Most of you have often experienced the specific phrase from the title of this chapter: think of the friend, and the friend appears, or rather think of the devil and the devil appears. While you are thinking of your friend, suddenly quite out of the blue your smartphone dings and you see that the friend is calling or has sent you a text. And then you keep asking yourself over and over again, how is it even possible!!! Are you some kind of "psychic" or something!!!

Well that's what this chapter is going to elaborate on. Such a telepathic ability is one of the many marvels of human brain. It's a kind of Extrasensory Perception (ESP) which is an intricate neurological characteristic of every human brain. It is the product of millions of years long evolution. Earth-Brain Bondage (which is discussed in chapter 6) and your love and affection towards the friend make this extraordinary cognitive skill possible. In technical term it is called as "psi phenomenon". Throughout history there have been many cases of subjective experiences of telepathy or clairvoyance (T-C) involving reminiscence, death, sickness or crises of friends, couples and relatives. When T-C experiences occur, they usually involve extreme emotional bondage between two human beings.

Human brain is a living transceiver, which is able to catch the emotional signals of another fellow brain. But how exactly the cerebral signals reach another brain irrespective of the distance in between. To explain this phenomenon, we have been developing hypothesis that ELF (extremely low frequency) electromagnetic fields are associated with T-C experiences. The ELF hypothesis is actually a generic label. In fact there are two most fundamental ELF fields that count for our extraordinary telepathic abilities, they are the Schumann's resonances (SR) and Geomagnetic pulsations. Since the time when human brain evolved, it has been immersed in these fields all along. Features of the Schumann resonance and geomagnetic field have been explained in chapter 6. The SR is responsible for relentless connectivity between two emotionally related minds or more specifically two cerebral biomagnetic fields. Geomagnetically quieter days allow the SR field to make the transference between two human brains more intense than the days with geomagnetic disturbances.

In a study done in 1980s, it clearly shows that spontaneous subjective T-C experiences are more likely to occur on days when the geomagnetic activity is lower (quieter) than on days before or after the experiences.

This pattern was statistically significant and was evident in the experiences that occurred between 1920 and 1967.

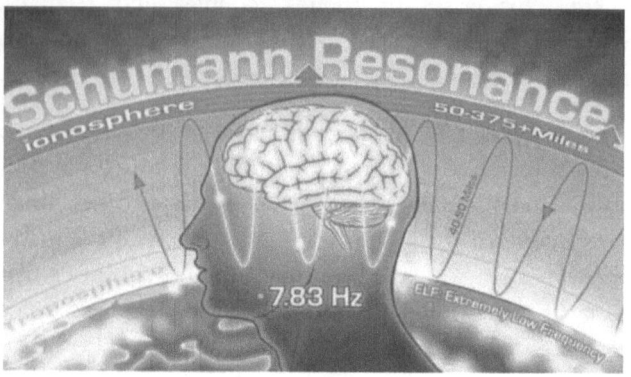

Figure 2.1 Schumann Resonance and Human Cerebral Field

So you can see that you are not the only person who may be experiencing such telepathic abilities. The strongest connectivity of minds is seen in mother-child relationship and in couples. For example, a mother is often able to feel the crises of her child who might be thousands of miles away. The same happens to couples who are deeply in love. If you are in love and you think that you can literally read your partner's mind before he or she even expresses something, then you really are a psychic with increased cerebral activity. With the assistance of the surrounding environmental electromagnetic fields the minds on planet earth remain connected forever.

But how exactly is it possible for a human mind to interact with another, through the carrier waves of environmental electromagnetic field! You will find in the later chapter Earth-Brain Bondage, human brain is able to feel, whatever our mother earth feels. These two systems share remarkably similar physical characteristics such as dynamic amplitudes and intrinsic frequencies. Therefore the possibility for direct interaction emerges. From this we can deduce that impedances and reluctances match or shared resonance occurs between the electromagnetic energies. The thin shell between the ionosphere and earth generates continuous harmonics of frequencies from a fundamental of about 7 to 8 Hz that is caused by global lightning which occurs between 40 to 100 times per second (40 to 100 Hz). Those frequencies are the Schumann resonances.

The bulk velocity of neuronal activity around the human cerebral cortices caused by the discharge of action potentials within this thin shell of tissue generates a resonance with a fundamental frequency between 7 to 8 Hz. On the bases of the average durations of the travelling waves over the cerebral cortices the repetition rate and phase velocities are in the order of 40 to 80 times per second (40 to 80 Hz). The current density around the annulus of an axon associated with a single action potential is equivalent to about 10^5 $A \cdot m^{-1}$ or the value associated with a single lightning discharge. For both cerebral and earth-ionosphere phenomena the average potential difference for these time-varying processes is in the range of 0.5 $mV \cdot m^{-1}$. The magnetic field component is in the order of 2 pT (10^{-12} T). The ratio of this voltage gradient to the magnetic field intensity is effectively the velocity of light, 300,000 km·s⁻¹.

Such quantitative evidence implies that the electric and magnetic fields of the Schumann resonances and those equivalent frequencies generated by the human cerebral cortices can interact persistently in a state of geophysical equilibrium or to speak simply, the communication occurs persistently in geomagnetically quieter days.

Now let's look into the mathematical aspect of the interaction between human mind and our planet. If you are not really into mathematics, you can just skip this part. A recent discussion of the Schumann Resonance characteristics summarizes that, the harmonics or modes of the Schumann resonances peak around 7.8 Hz, 14.1 Hz, 20.3 Hz, 26.4 Hz, and 32.5 Hz. This serial shift of about 6 Hz is consistent with the relation of:

$$[\sqrt{(n(n+1))}] \cdot [v \cdot (2\pi r)^{-1}]$$

where n are serial integers ≥ 1, v is the velocity of light in the medium (which is effectively c) and r is the radius of the earth. The first component of the relation when "n" is substituted as a quantum number is also employed to calculate the magnitude of the orbital angular momentum. When magnetic moments are expressed as Bohr magnetons the electron state is associated with a magnetic moment equal to $\sqrt{(n(n+1))}$. Changes in electron states or different shells are associated with emissions of photons. The peak values are not precise and can vary by ± 0.2 Hz depending upon ionosphere-earth conditions, time of day, season, influx of protons from solar events, pre-earthquake conditions, and yet to be identified sources. According to Nickolaenko and Hayakawa monthly variations in the first mode (7.8 Hz) range between 7.8 and 8.0 Hz. The diurnal variation in frequency shift has been attributed to drifts in global lightning or alterations in ionospheric height. Peak frequencies and amplitudes occur in May whereas minima occur in October-November. The peak to peak modulations are about ±25 % of the median value. The optimal metaphor is that every lighting strike of the approximately 40 to 100 per second between the ionosphere and ground is an expanding wave that moves until it ultimately interfaces with itself on the earth's spherical surface. The resulting interface elicits a return wave that arrives at the original source within about 125 ms after the initiation. The approximate equivalent frequency is 8 Hz.

The human cerebrum (1.35 kg) is an ellipsoid aggregate of matter that occupies about $1.3 \cdot 10^{-3}$ m^3. The three-dimensional metrics are: length (155-190 mm), width (131-141 mm), depth (108-117 mm). The cerebral cortices are approximately 1 to 4 mm thick but occupy almost 40 % of the cerebral volume with an average value of 490 cc. The surface area of the human brain is not smooth but exhibits convexities (gyri) and concavities (sulci). Two-thirds of the surface is buried within the sulci. Mathematical modeling indicates that this topological surface is similar to that of a flat surface "wrinkled" into the third dimension.

The primary source of the electromagnetic activity measured from the scalp emerges from the cerebral cortices because of the parallel arrangement of the dendrite-soma-axon orientations perpendicular to the surface for most of the approximately 20 billion neurons. The resulting steady state potential between the cerebral surface and a relative reference such as the lateral ventricle ranges between 10 and 20 mV. There is almost a linear correlation between the emergence of neuronal processes, and the magnitude of that potential. Superimposed upon the steady potential are fluctuating voltages that define the electroencephalogram (EEG). Most of the discernable frequencies occur within the ELF range of 1 and 100 Hz. However, fast frequencies up to 300 to 400 Hz, approaching the absolute refractory period of an axon, have been measured in epileptic brains.

The amplitudes of the scalp EEG range between 10 and 100 μV. In comparison corticographic discharges exhibit amplitudes between 0.5 and 1.5 mV. The most prominent dynamic pattern is the alpha rhythm (8 to 13 Hz) which during wakefulness is more evident over the posterior cerebral space with amplitudes in most (68 %) of the population between 20 and 60 μV.

Theta rhythms (4-8 Hz) have been considered "intermediate waves" that are significantly involved with processes associated with infancy and childhood as well as drowsiness and specific stages of sleep. The power of theta activity is evident even within the third decade (25 to 30 years of age) when the EEG parameters asymptote. This inflection is sometimes considered an index of cerebral maturation.

The other classical frequency bands associated with the human EEG are delta (1-4 Hz), which are the highest amplitude time-variations associated with Stage IV sleep, beta (13- 30 Hz) and gamma (30 to 50 Hz) patterns. The ranges are effectively arbitrary and related to EEG features associated with particular behaviors or power densities.

So, the fundamental brainwave EEG happens to fall within the basic harmonics of the Schumann Resonance peak around 7.8 Hz, 14.1 Hz, 20.3 Hz, 26.4 Hz, and 32.5 Hz. The typical strength of the electric field (SR peak harmonics) component is in the order of $mV \cdot m^{-1}$ while the magnetic field component is $\sim 1 \ pT \cdot Hz^{-1/2}$. Compared to the troughs of about 1 pT the peak intensities at the first and second harmonics are ~ 3 pT which decline to ~ 1 pT around 20.3 Hz and <0.5 pT at higher harmonics. The fundamental frequency and intensity increase by $\sim .04$ to .12 Hz and 0.11 to 0.41 pT, respectively during strong proton events (solar storms). The coefficients of this magnitude are slightly but significantly different along the X,Y,Z axes of propagation. Although these values appear minute, their potential can be realized by the magnetic energy from these values, described by:

$$J = B^2 (2\mu)^{-1} \ m^3$$

where B is the magnetic field strength, μ is magnetic permeability $(4\pi \cdot 10^{-7} \ N \cdot A^{-2})$, and m^3 is volume of the human cerebrum $(\sim 1.3 \cdot 10^{-3} \ m^3)$. The solution is within the range of $\sim 10^{-20}$ J. Just like the Unified Field Theory in modern physics, many of us neuroscientists are in an effort to come up with a Unified Theory for Neuroscience. The neuroquantum unit 10^{-20} J is the first step towards the Unified Theory. This quantum unit of energy is associated with the effect of the $1.2 \cdot 10^{-1}$ V net change during an action potential of a neuron upon a unit charge ($1.6 \cdot 10^{-19}$ A·s). That the activity of one neuron can affect the state of the entire cortical manifold of the brain has been shown experimentally.

In simple words, this single quantum of energy (10^{-20} J) from the Cerebral Electromagnetic field of thoughts and emotions would have the potential to disperse through the Schumann resonance ELF and reach another Cerebral Field thousands of miles away. In geomagnetically quieter days the environmental electromagnetic field carries the potential information of the specific thought or emotional response to the other brain and shares it with the cerebral field.

But when there is disturbance in the geomagnetic field the inter-cerebral signals get disrupted, just like during thunderstorm sometimes your cell network gets lost. Geomagnetic disturbance occurs due to different terrestrial and cosmological reasons, like solar storm (Coronal Mass Ejection CME), cosmic rays, lunar gravity interaction during full moon.

The cerebral fields of two emotionally involved human brains attain tuned brain wave EEG through the persistent Schumann resonance field which is in touch with both the brains at any specific moment, which means the more you get emotionally involved with your loved one, the more you'll be able to read each other's minds.

Once the minds are truly engaged, the next step would be to feel the other's physiological state of being. So, it is scientifically true that: *when you love someone deeply, every single cell in your body starts reacting when that special person is in distress.*

* * *

CHAPTER 3

SCIENCE OF EMPATHY & LEARNING - THE MIRROR NEURONS

Empathy is one of the various precious human characteristics that allows the humans as a species to connect with each other or even other creatures of earth. It is a bonding-mechanism between the living beings on our planet. Thanks to this amazing quality, you are even sometimes able to feel the pain and emotions of others. But how does this really work scientifically! Well… behind the curtain it's the work of another neurological wonder called the Mirror Neurons.

Most of you readers can already make sense out of the name itself. It works as a neurological mirror which reflects the emotional response through you from the person in front. Which means, when you see a couple hugging and kissing in front of you, don't be amazed if you start feeling the warmth of your beloved one's passionate embrace. As humans, we are evolutionarily designed to empathize with the other and share his/her grief or happiness even if that person is a complete stranger.

These marvelous cells of the human brain have literally shaped our civilization and these tiny wonders within our brain, make us aware of what it's like to be human. They are also the reason why you start yawning or giggling merely by watching another person do the same. They play the key role in a child's brain, while it is learning its mother tongue along with other cultural and sociological tactics. And as for adults, mirror neurons are your wingmen who help you in learning new skills. In the 1980s and 1990s, a few researchers Giacomo Rizzolatti, Giuseppe Di Pellegrino, Luciano Fadiga, Leonardo Fogassi, and Vittorio Gallese at the University of Parma, Italy discovered the mirror neurons in macaque monkey. The discovery was initially sent to Nature but was rejected at that time due to "lack of general interest", only to be accepted widely in a few years.

Mirror neurons were first found in various regions of the monkey brain (Macaca nemestrina and Macaca mulatta). So far scientists have observed the wonderful neurons in areas like the ventral premotor cortex (vPMC), inferior parietal lobe (IPL), primary motor cortex and dorsal premotor cortex (dPMC). Originally it was discovered, that the mirror neurons (MN) discharge both when the monkey does a particular action and when it observes another individual (monkey or human) doing a similar action. The name itself implies the significant feature of MNs. This specific feature of 'mirroring' or more specifically 'imitating' has been evolutionarily crucial in shaping the modern human civilization.

Figure 3.1 Mirror Neuron activity in the premotor cortex and parietal lobe of the monkey brain, when it grasps the banana as well as when it observes another individual grasping it

Neurophysiological experiments demonstrate that when humans observe an action done by another individual their motor cortex becomes active even without the presence of any motor activity. On this matter the first evidence was provided in the 1950s by Gastaut and his coworkers. They recorded significant changes (desynchronization) in an EEG rhythm (the mu rhythm) of humans both during active movements of studied subjects and when the subjects observed actions done by others.

Later many other researchers replicated the experiment and confirmed the results. The desynchronization while observing an action carried out by others includes rhythms originating from the cortex inside the central sulcus. Transcranial magnetic stimulation (TMS) studies provide us even more direct evidence to the existence of mirroring properties in the motor system of humans. TMS is a non-invasive technique for electrical stimulation of the nervous system. When TMS is applied to the motor cortex, at appropriate stimulation intensity, motor-evoked potentials (MEPs) can be recorded from contralateral extremity muscles. By modulating the amplitude of the MEPs through behavioral stimulation, it is possible to assess the significant effects of various experimental conditions. This way we can study the mirror neuron system in humans.

In the year 1995 Fadiga et al recorded MEPs, elicited by stimulation of the left motor cortex, from the right hand and arm muscles in volunteers required to observe an experimenter grasping objects (transitive hand actions) or performing meaningless arm gestures (intransitive arm movements). The results showed that the observation of both transitive and intransitive actions made a huge impact over the MEPs and increased the recorded MEPs. The increase concerned selectively those muscles that the participants use for carrying out the exact observed actions. Amplification of the MEPs while observing actions done by the others may result from the activation of the primary motor cortex owing to mirror activity of the premotor areas.

Later in the year 2000 another study by Strafella & Paus came up with support for this cortical hypothesis. By using a double-pulse TMS technique, they demonstrated that the duration of intracortical recurrent inhibition, occurring during the observation of an action, closely corresponds to that occurring during the execution of that specific action.

Another question rose from these studies, "Does the observation of actions done by others influence the spinal cord excitability?" In 2001 Baldissera and colleagues investigated the issue and discovered that there is an inhibitory mechanism in the spinal cord that prevents the execution of an observed action and leaves the motor cortex free to react to that action without the risk of any kind of movement generation.

Later many other neuroscientists showed that the motor cortical excitability faithfully follows the grasping movement phases of the observed action, or to say simply, your motor cortex area of the brain lights up whenever you see another person carrying out an action as well as when you carry it out yourself.

In conclusion, TMS studies indicate that a mirror-neuron system exists in humans and that it possesses important properties not observed in monkeys. Intransitive meaningless movements produce mirror neuron system activation in humans but not in monkeys. These properties of the human mirror neuron system (MNS) play an important role in determining the humans' capacity to imitate others' action as well as empathizing with them.

The Mirror Neuron System has vast impact over a person's social and behavioral skills throughout the lifetime. It allows us to be human and understand another human being and even other species for that matter. When you see a person get beaten up in the park, you suddenly start to feel his agony. The same happens when you see a street dog getting hurt. Humans are biologically designed to truly understand another creature's pain, happiness and desires, as if it is our own pain, happiness and desires. Feeling other's emotions and imitation learning are the most influential features of the mirror neurons.

Imitation is a great and the best mechanism of learning new skills in children and adult alike, although it is the most widely used form of learning during development, offering the acquisition of many skills without the time-consuming process of trial-and-error learning. Imitation is also central to the development of fundamental social skills such as reading facial and other body gestures and for understanding the goals, intentions and desires of other people. Many of us neuroscientists believe that there's a strong possibility that malfunction in the imitation learning mechanism or more specifically in the MNS may underlie various cognitive disorders, especially Autism.

During the developmental age of a child it learns language and different social skills. At this crucial stage of a person's lifetime, dysfunction in the mirror neurons might be one of the core deficits of socially isolating disorders such as Autism. A recent study by Iacoboni and colleagues highlights the importance of mirror neurons and their role in the development of autism spectrum disorder (ASD).

ASD is a pervasive developmental disorder characterized by impaired social interactions.

Iacoboni's team used functional magnetic resonance imaging (fMRI) to investigate neural activity of 10 high-functioning children with ASD and 10 normally developing children as they observed and imitated facial emotional expressions. Although both groups performed the tasks equally well, children with autism showed reduced mirror neuron activity, particularly in the area of the inferior frontal gyrus. Moreover, the degree of reduction in mirror neuron activity in the children with autism correlated with the severity of their symptoms.

These results indicate that a healthy mirror neuron system is crucial for normal social development. If you have broken mirrors or deficits in mirror neurons, you likely end up having social problems, as patients with autism do.

What about learning new skills in adults! Well… say you have moved as a professional to Paris, the most romantic city in the world with your beloved one. Although, you will be able to communicate with locals of Paris in English, but in order to look inside a Frenchman's inner self and understand his intentions, you'd require to learn the language of Paris' heart. In that case the neural basis of learning the sexy French, would be to listen to a lot of Frenchmen speaking French.

While your brain observes others speaking a foreign language it'd start firing in the Mirror Neuron Network and try to make sense out of it, by making you absorbed in the language either consciously or subconsciously.

The MNS would make you feel as if you are speaking the language in your mind. Eventually as you keep on listening to French along with a little literary assistance, the language centers of the brain; the Broca's area and Wernicke's area would start forming new neural network responsible for communicating in French.

Likewise, this imitation learning mechanism is responsible for enabling you to learn any new skills, like music, dance, mathematics etc. The more you observe and practice a specific skill, the MNS would fire more and the responsible area of the brain forms new neural pattern. This way you become better and better at a specific skill.

Humans are intensely social creatures. We share this feature with many other species. A complexity and sophistication that we do not observe among ants, bees or wolves, however, characteristically define the social life of primates. This complexity and sophistication is epitomized at its highest level by the social rules. Living in a complex society requires individuals to develop cognitive skills enabling them to cope with other individuals' emotions, intentions and actions, by recognizing them, understanding them, and reacting appropriately to them.

In a study, Singer and her colleagues used functional magnetic resonance imaging (fMRI) to measure brain activity in volunteers who observed others receiving painful stimulation to their hands. As expected, mere observation of another's pain produced increased activation in the pain center of the observer's brain, including the insula and anterior cingulate cortex.

In the earlier study of Singer et al. (2004), people who scored higher on standard empathy scales had higher activity in these brain areas. It thus appears that more empathetic people have more active mirror neuron systems for appreciating the pain of others.

The major manipulation of the Singer et al. (2006) study was that the people who received painful stimulation had previously engaged in a game where some had behaved fairly and others unfairly. Men, and to a lesser extent women, showed much less pain center activation in the brain for those sufferers who had acted unfairly. Moreover, men but not women showed greater activation in the reward center or pleasure center of the brain, the nucleus accumbens when observing unfair people being punished. Thus men more than women took pleasure in the pain of wrongdoers.

We literally live by emotions. Knowingly or unknowingly we always get triggered by another person's intentions and feelings. Mirror Neuron System makes it possible for us to empathize with a fellow human being. That's why movie stars are so good at making the viewer burst into tears over a scene.

When the actors put all their excellence and effort in making the characters real, the viewer brain starts to feel the character's suffering. You start feeling as if you are the character in the movie yourself. Which means, while watching Casablanca, for a few hours you become Humphrey Bogart or Ingrid Bergman yourself. Hence, the Mirror Neuron System acts as a bonding agent among humans through the mechanism of empathy.

* * *

Chapter 4
Meaning of Dreams

Dreams are the gateway towards one's soul. Many consider them as a path to the other world, the supernatural kind. And even, throughout human history many people have received vision of God or commands from the Supreme Soul in dreams. Many even see their dead son, wife, husband or other dead relatives in their dreams. Even great scientists have perceived scientific revelation in their dreams. As if the cosmic record of information opens up to them through their dreams.

But, are there really any scientific basis to all these experiences! Actually YES... all these can be explained without involving the supernatural component. Dream-Interpretation is one of psychology's most celebrated matters of interest.

Basically dreams are brain's own mechanism of getting rid of unnecessary information throughout the daytime. It's a process of reverse memory. Often they also produce a virtual world of wish-fulfillment, where the most intense wish of yours comes true. This is just another way to keep you satisfied with the virtual simulation of the things you crave for. The world of dreams is really a mysterious and wonderful one. Let's explore the world.

Dreaming is a byproduct of REM (Rapid Eye Movement) sleep, while the brain works on the daily calibration. But first let's answer the question what a dream really is!

Well... dream refers to the subjective conscious experiences we have during sleep. We neuroscientists define dream as a subjective experience during sleep, consisting of complex and organized images that show temporal progression of visual imageries. Dreaming is a universal feature of human experience.

Why do we have vivid, intense, and eventful experiences while we are completely unaware of the world that physically surrounds us? Couldn't we just as well pass the night completely unconscious? Not really. The function of dreaming has remained a persistent mystery for a long time, but with the discovery of various active and inactive brain regions during REM dream state we are finding out the impact of dreams over the awake state of mind.

Cognitive neuroscientists and psychologists throughout the human history have put forward various theories on the function for dreaming. Dream consciousness is perceived as some kind of noise generated by the sleeping brain as it fulfills various neurophysiological functions during REM sleep. But don't even for a second think that this noise is completely meaningless. This is one of the most important meaningful noises in the world of biology.

In many cultures dreams have been hailed as messages from the gods and dismissed as random hallucinations. The pendulum of popular opinion has swung from one extreme to the other throughout recorded history and between cultures and camps, with scientists, psychologists, sages, and philosophers all weighing in. Aristotle, for one, believed dreams were formed by the dreamer's impaired mind, and Plato argued that dreams represent a frightening breakdown of reason.

In the Victorian era, some scientists posited that dreaming was pathological. But rather than placing dreaming on a mystical pedestal, or looking at dreaming as a deficient form of consciousness, it is highly instructive to look at dreaming as an alternative form of consciousness and a different way of thinking.

Though Plato and Aristotle could not have proven it, today we know that the dreaming brain is, in a sense, differently abled. At least two important regions, the dorsolateral prefrontal cortex (DLPFC) and the precuneus in the parietal lobe, are deactivated during rapid eye movement (REM) sleep, the period when most dreaming takes place. Because of this, we lack the ability to fully exercise our short-term memory when we dream, both within the dream and upon awakening. This helps explain breaks in continuity during the dream and why it is difficult to recall dreams on waking. Also, we are unable to locate our physical body in space when asleep, which is why the dreamer does not realize her or his body is at home in bed during nocturnal adventures in familiar or fantastical landscapes. Making decisions or directing will is likewise difficult while dreaming as the regions responsible for those cerebral activities are inactive during the REM sleep.

My own fascination with dreams began after studying a renowned work of Sigmund Freud called Interpreting Dreams. He was one of the early frontiers of psychology who dared to step into the mysterious arena of dream-analysis. He is known as the father of psychoanalysis. But Freud had a major flaw in his works. He presumed that all behavioral cerebral activities have something to do with libido.

Just imagine, according to Freud all your activities are connected to sexuality. Come on... let's be honest. It doesn't take a neuroscientist to figure out that it's not true. Sorry Freud, your idea of sex drive driving all brain functions doesn't hold water. But Freud did have some really intriguing functional theories about dreams. One of the most important of Freudean ideas about dreaming was wish-fulfillment. Sigmund Freud and Carl Jung mentioned in their works that *"dreams are the royal road to the unconscious"* and *"the most readily accessible expression of the unconscious"*. I'll elaborate on that in a while.

But first let's look into the neurobiology of dreaming. To understand dreams first you need to understand various stages of sleep in which dreams occur. Previously experts divided sleep into five different stages. Fairly recently stage 3 and 4 have been combined, so in your sleep you usually pass through four stages: 1, 2, 3 and REM sleep. These stages progress cyclically from 1 through REM then begin again with stage 1. Stages 1 to 3 collectively is known as Non-REM sleep. A complete sleep cycle takes an average of 90 to 110 minutes.

Now the question may rise in some of your minds is that what are REM and NREM sleep? Most of you may have already heard about REM sleep and Non-REM sleep. Rapid Eye Movement sleep is a very crucial stage of sleep characterized by the rapid and random movement of the eyes.

In 1953, Aserinsky and Kleitman discovered human rapid eye movement (REM) sleep and documented that dream reports were obtained most frequently when subjects were awakened from REM sleep. People with suspected sleep disorders lack REM sleep. Mostly dreaming occurs in REM sleep, but here is a long-standing controversy surrounding the existence of dream experiences during non-rapid eye movement (NREM) sleep. Dream reports from NREM sleep were less remarkable in quantity, vividness and emotion than those from REM sleep.

NREM sleep dreams are more frequent during the morning hours when the occurrences of REM sleep are highest. 80% of your sleep is NREM sleep while the rest 20% is REM sleep. There is little or no eye movement during NREM sleep. Each stage of NREM sleep lasts around 5-15 minutes.

Stage 1: Your eyes are closed, but it's easy to wake you up.

Stage 2: You are in light sleep. Your heart rate slows and your body temperature drops. Your body is getting ready for deep sleep.

Stages 3: This is the deep sleep stage. It's harder to rouse you during this stage, and if someone wakes you up, you would feel disoriented for a few minutes.

During the deep stage of NREM sleep, the body repairs and regrows tissues, builds bone and muscle, and strengthens the immune system. As you get older, you sleep more lightly and get less deep sleep. Aging is also linked to shorter time spans of sleep, although studies show you still need as much sleep as when you were younger. Which means that even though the sleep gets lighter with age, try to get as much sleep as you used to have during your youth in order to wake up rejuvenated.

REM sleep stage is the most crucial stage for dreaming. Newborns spend around half of their sleep in REM stage, which helps in the development of the baby brain. And studies have shown that REM sleep is very much important in learning new skills at all ages. One study found that REM sleep effects learning of certain mental skills. People taught a skill and then deprived of NREM sleep could recall what they had learned after sleeping, while people deprived of REM sleep could not.

Going deeper in the mysterious and captivating world of dreams we have found various regions of the brain that are active and inactive during REM sleep. While a great deal of our brain remains active in REM sleep, regions related to executive functions such as rational thought, linear logic and episodic memory, as well as primary sensory and motor functions remain relatively inactive or just simply asleep. Just imagine, all the regions that enable you to make sense of various events, are literally turned off. Then how can it be possible for the events or progression of imageries to be logical or real!!! All the imageries perceived during REM sleep are merely the by-products of random neural firings, while the brain turns on the calibration in order to be ready and refreshed for the next day activities.

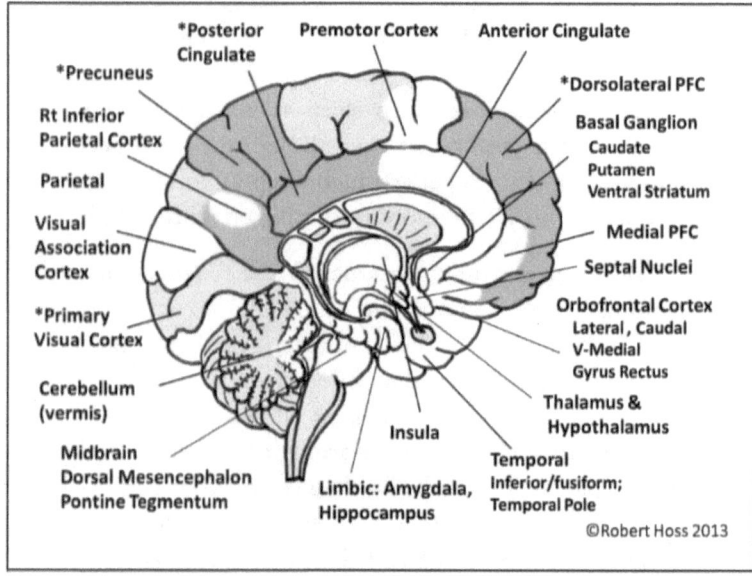

Figure 4.1 Relatively active (white) and *inactive (dark gray) centers of the brain in REM sleep. Derived from neuroimaging studies (Maquet, Braun, Nofzinger et. al. in Hobson 2003)

A number of active regions appear to be involved in the perception of the dream. Although the primary visual cortex and much of the parietal cortex remain inactive, activity is heightened in the visual association cortex, which processes imagery associations, and the right inferior parietal region, which organizes the imagery into a visual space.

Other sensory dream experiences may be due to internally stimulated activity in the vermis cerebellum and other motor and sensory regions as well as activity in the temporal areas involved in facial recognition, auditory processing and episodic recall.

Activity in the visual association regions of the cortex gives rise to the picture-metaphor nature of dream imagery, i.e. picturing the emotional, memory and conceptual associations.

This explains the visual imagery you have of your dearest one in your dreams. Here Freud's theory of wish-fulfillment proves its mettle. As an evolutionary advantage, your brain literally fulfills your most intimate wishes in your dreams and makes a virtual reality out of it, which doesn't seem any different than the real thing. While you dream of your beloved one whether it is your wife, husband, lover, son or daughter, in your dreams you truly feel the oneness with them. In many cases it may not be a person people dream about, rather an object they wish to have. So if you crave for something or someone strongly enough, your dreams will give you the satisfaction of having them in a virtual reality, if not in real.

The majority of the brain centers which are active during REM dreaming are those which process mental material either unconsciously or prior to their actions becoming conscious. Dreams therefore provide us with valuable access to the unconscious. This proves that Freud and Jung were right about dream being the royal road to the unconscious. There are a number of cognitive centers in the frontal regions of the brain that are highly active in REM sleep. This suggests that the dreaming brain is capable of problem resolution and creative insight. For scientists who are truly absorbed in their work, even their unconscious mind keeps solving problems when they are having REM sleep. As a result, they often find solutions related to their work which they usually cannot find while being awake.

Figure 4.2 The Great Autodidact Mathematician Srinivasa Ramanujan

The great mathematician Srinivasa Ramanujan had visions of many of his mathematical equations in dreams. Most of the time in his dreams he used to see that a hand is writing various strange equations on a wall. Suddenly he used to wake up and start working on the dream imageries of those equations.

The creative problem solving history of dreams is well documented by Barrett (2001, 2007) who researched the many inventions and artistic creations arising from dreams. She describes dreaming as *"thinking in different biochemical state"*. Her research shows that anything may get solved during dreaming, particularly if the problem involves visualization or where the solution lies in "thinking outside the box".

But in most of the dreams people have, they see the metaphoric representation of their dynamic life situations. The metaphor of dreams becomes more obvious and eloquent when the dreamer tries to translate the dreams in his own words. The dream finds its own way of interpretation when it gets access to the dreamer's language center of the brain.

Let me give you an example; a man who had become frustrated and miserable at work but was not sure why, describes his dream:

"I had a frightening dream where I was being chased away by a big buffalo with a little buffalo following it."

When he was asked what a big buffalo does, he said:

"he is huge and powerful, when he wants you to go, you go,"

which he recognized as describing his boss. Then he was asked about the little buffalo. He said,

"He is a little pipsqueak that follows the big one around -- just like that little pipsqueak at work!"

The dream revealed that the source of his restlessness was not only the actions of his boss but the relationship that this little pipsqueak of a co-worker had with his boss. Here the metaphors aptly pictured the emotional similarities between the big buffalo and his boss, and the little buffalo and the 'little pipsqueak' at work.

Dreams are intensely influenced by emotions. The limbic areas, in particularly the amygdala, is highly active during the REM dream state. Dream creates imageries based on your emotional inclination. Emotion doesn't arise from the dreams, rather your deepest emotion orchestrates your dream. Many times your emotional feel-good state or frustrated state is woven into dreams with the use of picture-metaphors. A woman, who typically felt in control of the events in her life, suddenly learned from the doctor that her husband was terminally ill and she could do nothing about it. That night she dreamed,

"I was locked in a car with no steering wheel and no door handles or window controls. I was rolling backward down a steep hill, and there was no way of stopping it, or getting out of it. I woke up in a panic."

Here the feeling of being totally unable to control the situation was pictured with all of its emotional intensity and vividness.

So as you can see, dreams are not just meaningless random neural firings of the brain. They really are the reflection of your unconscious mind. Basically we humans are emotional species. And this emotional trait even gets reflected in the dreams. High activity in the amygdala and associated limbic system has led us to conclude that dreams selectively process emotional memories through the interplay between the cortex and the limbic system.

So for now we can say that the amygdala actually orchestrates the dream activity. The virtual simulation of all emotional responses in your dreams in fact makes the brain gain more control over real life situations. Dreams have the potential to enable humans to handle even stressful circumstances smoothly.

Freud suggested that bad dreams let the brain learn to gain control over emotions resulting from distressing experiences. Studies have shown that dreaming metaphorically represents projections of emotional expectation and in many cases lowers stress levels (due to a massive reduction in stress producing neurotransmitters norepinephrine in forebrain centers including the amygdala).

Likewise nightmares are extreme form of emotional processing in dreams. But remember, people usually term every bad dream as a nightmare, while scientifically a dream is not a nightmare unless it is extremely upsetting, containing overwhelming anxiety, apprehension and fear. Nightmares can have a number of causes including heavy emotional stress, severe threat to self or self-image, unresolved or extreme trauma (PTSD), psychological problems, the influence of certain drugs, emerging medical problem requiring attention, or sleep disorders affecting REM/NREM sleep balance. Resolving the underlying cause can lead to the freedom from nightmares.

But not always dreams are so obvious and easily interpretable. Sometimes dreams are exquisitely bizarre that really just don't make any sense. In that case, you don't need to be confused, rather calm your mind and let the dream itself find its way through speech. And while you start translating it, you'd be amazed to see that you can interpret the dream and make proper associations between the dream content and your life events.

* * *

CHAPTER 5

STUDIES IN HYSTERIA - EMOTIONAL SUPPRESSION AND ITS DANGEROUS CONSEQUENCES

Are you a kind of person who likes to keep all your emotions hidden from the people around you! Do you prefer restraining your feelings a little too much! In that case, you must know that too much emotional suppression can have catastrophic impact over your body. The results are blindness, paralysis, numbness and other major neurological problems. This kind of physiological condition caused by emotional suppression is known as Conversion Disorder and the older term for it was "Hysteria". Now the word hysteria is commonly used to describe unmanageable emotional excess.

We get to know the essentials of hysteria from the joint work of Sigmund Freud and Josef Breuer, known as "Studien uber Hysterie" (Studies in Hysteria), first published in 1895. The book consists an introductory paper and five individual studies of hysterics (Breuer's famous case of Anna O. and four more by Freud). Before Freud and Breuer, it was the French neurologist, Jean-Martin Charcot who brought the attention of the world towards hypnosis and hysteria. He initially believed that hysteria was a neurological disorder for which patients were pre-disposed by hereditary features of their nervous system, but near the end of his life concluded that hysteria was a psychological disease. But the condition of Conversion Disorder or Hysteria dates back to ancient Greece.

Some doctors falsely believe that this disorder is not a real condition as it doesn't have any underlying biological cause, so they abruptly tell their patients that the problems do not exist in real, it's all in the patient's head. People with Conversion Disorder are not faking (malingering) their symptoms, although there is a syndrome in which a person feigns disease, illness or psychological trauma to draw attention and sympathy; it is known as Munchausen Syndrome.

Conversion disorder (functional neurological symptom disorder) is classified as one of the somatic symptom and related disorders in the Diagnostic and Statistical Manual of Mental Disorders of the American Psychiatric Association, Fifth Edition, (DSM-5). This can happen to anybody. The harder you try to keep your feelings inside, which you think as inappropriate to express, the more vulnerable you become to Conversion Disorder symptoms. For example; in an argument you get so intensely mad at someone that you want to hit that person, but as you think it might be totally inappropriate, you restrain yourself and don't let the anger out. The result of such extreme suppression can be catastrophic.

In similar events, many people have lost their ability to speak or have even become paralyzed. Does that mean when you feel like hitting someone, you should definitely do that! Not at all… there is another way to deal with this kind of situation. You just need to talk your feelings out to a close friend. How you feel! How much rage is bursting inside you! How much you want to pull off the person's head and play basketball with it! Spit it all out.

In most cases Hysteria symptoms occur because of any psychological conflict. The emotional battle that you feel within, the conflict between guilt and evil, rage and peace can turn the world of physical body upside down. Although Freud and Breuer were not first ones to work on Hysteria, their work made a real impact over the scientific community.

While Breuer, with his intelligent and amorous patient Anna O., had unwittingly laid the groundwork for psychoanalysis, it was Freud who drew the consequences from Breuer's case. The psychological conflict can have various factors underlying.

On this matter Freud had a huge misperception. He wanted a grand unifying theory for all hysteria symptoms. That's why he always was obsessed with his theory of sexual conflict underlying hysteria. Breuer on the other hand kept on finding different factors such as extreme emotional suppression, varieties of trauma etc.

This can happen to anybody, but documented literature shows more number of women suffering from Conversion Disorder symptoms than men. But why exactly, women are more vulnerable to such symptoms! All psychological states are the product of various hormonal interaction within the body. And a female body goes through way more hormonal changes in her lifetime than a male body. So it is easier for a psychological conflict to influence the hormonal interaction within a woman's body than a man's.

Hippocrates (5th century BC) was the first to use the term hysteria. Indeed he also believed that the cause of this disease lies in the movement of the uterus ("hysteron"). The Greek physician provides a good description of hysteria, which is clearly distinguished from epilepsy. He emphasizes the difference between the compulsive movements of epilepsy, caused by a disorder of the brain, and those of hysteria due to the abnormal movements of the uterus in the body. Then, he resumes the idea of a restless and migratory uterus and identifies the cause of the indisposition as poisonous stagnant humors which, due to an inadequate sexual life, have never been expelled. He asserts that a woman's body is physiologically cold and wet and hence prone to putrefaction of the humors (as opposed to the dry and warm male body). For this reason, the uterus is prone to get sick, especially if it is deprived of the benefits arising from sex and procreation, which, widening a woman's canals, promote the cleansing of the body. And he goes further; especially in virgins, widows, single, or sterile women, this "bad" uterus, since it is not satisfied not only produces toxic fumes but also tends to wander around the body, causing various kinds of disorders such as anxiety, sense of suffocation, tremors, sometimes even convulsions and paralysis. For this reason, he suggests that even widows and unmarried women should get married and live a satisfactory sexual life within the bounds of marriage. However, when the disease is recognized, affected women are advised not only to partake in sexual activity, but also to cure themselves with acrid or fragrant fumigation of the face and genitals, to push the uterus back to its natural place inside the body.

Even Aristotle and Plato seemed to have perceived the symptoms of Hysteria as a result of a sad uterus. Somehow all these geniuses of history had the idea that hysteria was related to the lack of sexual pleasure. Plato, in Timaeus, argues that the uterus is sad and unfortunate when it does not join with the male and does not give rise to a new birth.

But don't worry, we are not going to mess with anybody's uterus. That might be considered as sexual harassment. So, my dear Hippocrates, Plato and Aristotle... thanks... but no thanks.

Modern neuroscience shows us that it's not really about whether the uterus is sad, rather it's about whether you are sad, and yet not sharing your sorrow with anyone. A sad uterus is just a symbolism for the dissatisfaction in the sexual life of a woman. This implies that sexual suppression may come out as physical problems, if you cross the line of repression. But again, sexual repression is just one among many factors in Conversion disorder symptoms. There are tons of other factors that can cause those symptoms. For example a sudden mental trauma can leave a person totally blind or mute.

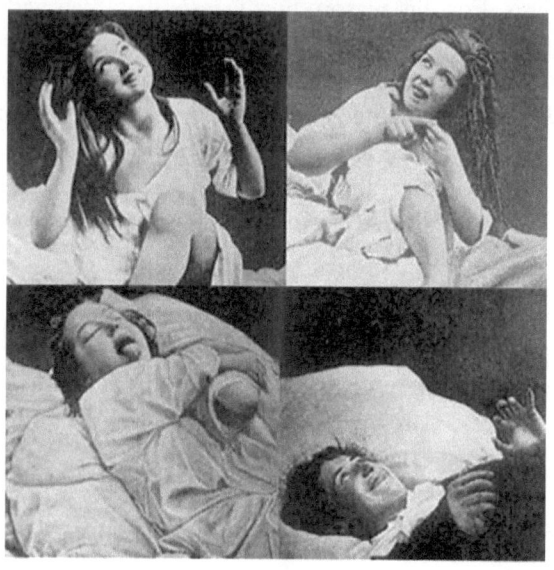

Figure 5.1 A Hysteria Patient Under The Effects of Hypnosis

French neuropsychiatrist Pierre Janet (1859-1947), with the sponsorship of J. M. Charcot, opened a laboratory in Paris' Salpetriere. He convinced doctors that hypnosis based on suggestion and dissociation was a very powerful model for investigation and therapy. He wrote that hysteria is *"the result of the very idea the patient has of his accident"*. The patient's own idea of pathology is translated into a physical disability. Janet studied five hysteria symptoms: anaesthesia, amnesia, abulia, motor control diseases and modification of character. She suggested that the reason of hysteria is in the *"idee fixe"*, that is the subconscious. Janet's studies are very important for the early theories of Freud, Breuer and Carl Jung.

The most interesting case in hysteria's history was Breuer's patient Bertha Pappenheim, alias Anna O. According to Breuer, Anna *"had hitherto been consistently healthy and had shown no signs of neurosis during her period of growth. She was markedly intelligent, with an astonishingly quick grasp of things and penetrating intuition. She had great poetic gifts, which were under the control of a sharp and critical common sense"*.

Figure 5.2 Breuer's Patient Anna O.

Breuer reported, Anna fell prey, during her father's final illness and in the months after his death, to the most appalling symptoms of hysterical paralysis and anaesthesia in three out of her four limbs, together with a succession of other distressing psychiatric symptoms.

At different times these included weakness, inability to turn her head, diplopia, a nervous cough, loss of appetite, hallucinations, agitation, mood swings, abusive and destructive behavior, amnesia, somnolence, tunnel vision and partial aphasia

Breuer further recorded: *"She no longer conjugated verbs and eventually she used only infinitives, for the most part incorrectly formed from weak past participles".* Among her symptoms, she was at one time unable to speak in her native German, but could still read both French and Italian, translating them aloud into English. On one occasion she was for several weeks unable to drink in spite of a tormenting thirst. Often she experienced alterations in her personality accompanied by confusion and delirium, this state was called "absence".

Breuer observed that, while the patient was in her state of absence, she was in the habit of muttering a few words to herself which seemed as though they arose from some train of thought that was occupying her mind. During part of her illness, she was unable to recognize or accept food from anyone except her physician, who spent almost a total of a thousand hours with her between April 1881 and June 1882. She was able to satisfy herself of his identity only by holding his hands.

Notice that all her symptoms were not the product of any organic cause, rather they were totally psychological. Breuer figured out that as Anna started to talk all about her fairy tales, previously repressed feelings and thoughts, her hysteria began to disappear. She coined the term "talking cure" for this kind of talking treatment. Sometimes jokingly she used to call this "chimney sweeping". Later these simple words of Anna O. got their sophisticated forms as "psychotherapy" or "counselling".

Each evening Breuer would return and Anna would recount, with vivid emotion, the exact events from precisely one year ago. In the final stage, Anna began to add to these accounts a description of the various occurrences that had evidently triggered each of her hysterical symptoms during the previous year. As she did so, the relevant symptom itself would disappear. For example, on recalling her disgust at seeing a dog drink from a lady companion's glass of water a year before, she was suddenly able to drink once more. She recovered from her symptoms over time, and later in life she became a distinguished social worker and a noted campaigner for women's rights.

But let's not use the term "Hysteria", as it is not the modern day term for the disorder. So, we are going to use the term Conversion Disorder while exploring the neurobiology behind it. Conversion disorder is a specific form of somatization in which the patient presents with symptoms and signs that are confined to the voluntary central nervous system.

Now, what is "Somatization"?

Somatization is the psychological mechanism whereby psychological distress is expressed in the form of physical symptoms. The psychological distress in somatization is most commonly caused by any kind of psychological conflict that threatens mental stability. Conversion disorder occurs when the somatic presentation involves any aspect of the central nervous system over which voluntary control is exercised. However no specific neurological pathway is discovered, that acts as a bridge between the psychological conflict and physical symptoms like Anaesthesia, Paralysis, Ataxia, Tremor, Partial Seizures, Deafness, Aphonia, Globus hystericus.

But still, several studies have given us amazing leads on the neurobiology of Conversion Disorder. As there is emotion involved in the Conversion symptoms, it is not farfetched to say that the original emotion center (presently we consider only the amygdala as the emotion center) of the brain, the limbic system is deeply involved.

Preliminary data from neuroimaging studies provide us evidence on possible networks engaging the limbic system and motor regions that may be involved in conversion symptoms like paralysis. Studies demonstrate the engagement of regions in the limbic-motor interface to attempted or imagined movement (ventromedial prefrontal cortex) and non-noxious brush stimuli (caudate/putamen) in conversion paralysis. These regions have been suggested as potential nodal points for emotional stimuli to influence motor function.

As a part of the limbic system, amygdala handles most of the emotional processes. There are several mechanisms by which amygdala activity may modulate motor behaviors. In a study by Valerie Voon et. al., they demonstrated aberrant limbic-motor interactions in patients with conversion disorder that may underlie the influence of affect or arousal on motor function.

Patients with motor conversion disorder had greater functional connectivity from the right amygdala to the right supplementary motor area, a region involved in motor initiation. Thus any imbalance in the emotional state can have potential impairing effects on the body. This explains how a trauma from a sudden death of the spouse can leave a person paralyzed or even catatonic.

So, the tagline is DO NOT SUPPRESS YOUR EMOTIONS TOO MUCH. You won't even notice when you just cross the line and step into the catastrophic domain of Conversion Disorder. It's better not to hold your feelings inside too much and express them to a dear one freely, than to pay thousands of dollars to a psychiatrist for the same outburst of emotions later. Emotions are a bonding mechanism for humans. So, use 'em, abuse 'em and utilize 'em.

* * *

CHAPTER 6

EARTH-BRAIN BONDAGE - PSYCHOLOGICAL & PHYSIOLOGICAL CHANGES DURING FLUCTUATIONS IN GEOPHYSICAL PARAMETERS

Mother earth has put all her excellence in molding the human brain throughout the millions of years long evolutionary period. Just like, a mother's genetic traits are forwarded to her children, our beloved mother earth with the use of Darwinian concept of adaptation and naturally selective pressure, has gifted us an amazing and marvelous neural network of the brain which is capable of contemplating the universe, the limitless cosmos, the beauty of mother nature, and even capable of contemplating itself contemplating. The beautiful human brain is even able to feel what our planet is feeling.

Have you ever thought, why and how patients suffering from schizophrenia and paranoia happen to have more psychotic breaks at specific periods of the lunar cycle! This happens due to the disturbance in the geomagnetic field of planet earth caused by lunar gravity. Likewise a wide variety of cerebral, biological responses and medical complications has been associated with geophysical events. These effects include alterations in occurrences of migraine headaches, strokes, glaucoma pains, joint swelling (objectively measured), blood clotting, skin conductivity, tissue permeability, embolism risks, thyroid activity, heart attacks etc. In this chapter I'll describe how deeply our beloved mother earth is ingrained within our brain or to talk scientifically, I'll explain the interconnectedness between the earth's geomagnetic activity and the cerebral activity. I'm about to show you the fantastic craftsmanship of planet earth while molding the 3 lbs. lump of jelly.

The human brain is perhaps the most intricate and advanced creation (not from a creationism perspective, but from a completely evolutionary) of Mother Nature on earth. At present times, the mesmerizing exploration of mysteries within the human brain is still in infancy. We have only started to take baby steps.

If we go deeper into the spectacular network of the human brain, all we shall experience is a feeling of awe. At every level of the mesmerizing neural circuitry we shall discover an undeniable signature of Mother Nature or more specifically planet earth. Humans have not just evolved on this planet, indeed, the planet molded the brain as per its own specific criteria and even sowed the seeds of its own characteristics deep within the neural network.

During a long period of 15 million years, since the time when Hominids (Great Apes) diverged from the Gibbon family, evolution has slowly made us everything we are. As a result of that prolonged process of biological evolution we see mysterious neurological conditions occurring during global or solar events, which only make sense when we take various geophysical and cosmological factors into account.

Human brain is the most complicated mystery of our planet. The womb of mother earth is filled with mysterious mechanisms. These mechanisms make the creatures of this planet uniquely special in one way or another. Every living organism on earth is always connected to the geophysical state of the planet. We see the beauty of this correlation everywhere around us, most apparently in animals. For example, an accurate navigation system has been gifted to various earthly creatures by our planet, which is called "Magnetoreception". It works in pace with the earth's geomagnetic field and helps the migratory birds to find their way back home.

To explain a little more we can say that the avian magnetic compass is a complex biological mechanism with many surprising properties. The basis for the magnetic sensing is located in the eye of the bird, and furthermore, it is light-dependent, i.e., a bird can only sense the magnetic field if certain wavelengths of light are available. Specifically, many studies have shown that birds can only orient if blue light is present.

Likewise, animal behavior prior to earthquakes is quite captivating to human imagination. All animals instinctively respond to escape from predators and to preserve their lives. A wide variety of vertebrates already expresses "early warning" behaviors right before an earthquake. So it's plausible that a seismic-escape response could have evolved from this already-existing genetic predisposal. An instinctive response following a Primary wave seconds before a larger Secondary wave is not a huge leap, so to speak, but what about other precursors that may occur days or weeks before an earthquake! In fact there are precursors to a significant earthquake that we have yet to learn about, such as ground tilting, groundwater changes, electrical or magnetic field variations. Indeed it's possible that some animals could sense these signals and connect the cognitive phenomenon with an impending earthquake.

Now let's focus on the main mettle of this chapter. What kind of earthly signature the human brain has, without even being aware of it. Homo sapiens brain has been immersed in the geomagnetic field of planet earth from the very beginning of human evolution. As a result every single neural firing in the synapses has been interacting with the surrounding geomagnetic field at all times. If we go back in time we'll find out that, a major adaptive advantage of human evolution was an amazing increase in brain size. Fossil evidence allows us to trace the development of the brain as it increased threefold over the last three million years. Throughout human evolution, the brain has continued to expand.

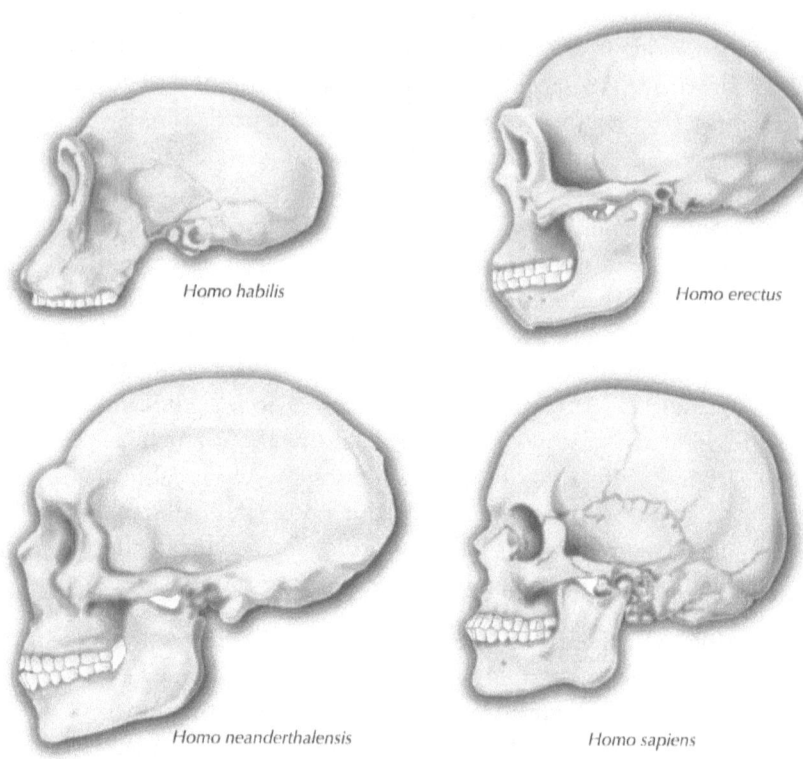

Figure 6.1 Series of hominid skulls with different brain sizes (Homo habilis 550 cc, Homo erectus 1100 cc, Homo neanderthalensis 1500 cc, Homo sapiens 1300 cc)

The earliest documented members of the genus Homo are Homo habilis, which evolved around 2.4 million years ago. This was the earliest species for which there is positive evidence of use of stone tools. The brains of these early hominins were about the same size as that of a chimpanzee, although it has been suggested that this was the time in which the human SRGAP2 gene doubled, producing a more rapid wiring of the frontal cortex, that is responsible for analytical thinking.

During the next million years a process of encephalization began, and with the arrival of Homo erectus in the fossil record, cranial capacity had doubled from 550 cc to around 1100 cc. This increase in human brain size is equivalent to every generation having an additional 125,000 neurons more than their parents.

Homo erectus were the first species in the history of planet earth that learnt to tame and create fire. They also excelled in making complex tools. Throughout this entire human evolution period planet earth had long enough time to mold the human brain and make it most suitable to survive and further evolve on this planet.

EEG pattern of the brain waves are the most eloquent evidence of planet earth's geomagnetic touch over cerebral activity. Lewis B. Hainsworth of Western Australia seems to be the first researcher to recognize the relationship of brain wave frequencies to the naturally circulating rhythmic signals of the planet, known as Schumann's Resonance (SR).

In 1952, a German physicist, Dr W.O. Schumann, suggested that the space between the surface of the earth and the ionosphere acts as a resonant cavity, somewhat like the chamber in a musical instrument. Energy for the SR is provided by lightning. Lightning pumps energy into the earth- ionosphere cavity and causes it to resonate at frequencies in the ELF range.

Schumann calculated the SR frequencies and fixed the most predominant standing wave at about 7.83 Hz. The frequency values of the SR signals are determined by the effective dimensions of the cavity between the Earth and ionosphere. Thus, any kind of event that changes these dimensions will change the resonant frequencies. Such events could be ionospheric storms, and could even result from a man-made ionospheric disturbance. These ionospheric disturbances whether they are man-made or caused by a solar-storm (Coronal Mass Ejection), produce significant geomagnetic storms and anomalies. These geomagnetic conditions change diurnally, seasonally and with variations in solar activity, which, in turn varies with the 11 year sunspot cycle and also with the 27-29 day lunar cycle, mainly during sunspot minimum periods.

Hence, these changes in the atmospheric conditions sometimes affect the biological systems. For instance, lunar cycle creates geomagnetic disturbance, which in turn proves to be the precursor and stimulant to rheumatoid arthritic pain in some aged people. A number of biologists have concluded that the frequency overlap of SR and biological fields is not accidental, but is the culmination of a close interplay between geomagnetic and biomagnetic fields over evolutionary time. Hence researchers have examined interactions between external fields and biological rhythms. Organisms are capable of sensing the intensity, polarity and direction of the geomagnetic field. A variety of behavioral disturbances in the human population is statistically related to disturbances in the geomagnetic field :

Friedman et al (1965) documented a relationship between increased geomagnetic activity and the rate of admission of patients to 35 psychiatric facilities.

Venkatraman (1976) and Rajaram & Mitra (1981) reported an association between changes in the geomagnetic field due to magnetic storms and frequency of seizures in epileptic patients.

Geophysical features determine the frequency spectrum of human brain wave rhythm. Moreover, the frequencies of naturally occurring electromagnetic signals, circulating in the electrically resonant cavity bounded by the Earth and the ionosphere, have governed or determined the evolution or development of the frequencies of operation of the principal human brain wave signals. Which means, the geophysical parameters nourished and developed the human brain, and in fact the entire anatomy, to live in the tuned system of planet earth by means of the naturally occurring Schumann's resonances.

This way, the magnetic and electric field strength of the human species have approached the values same as Schumann's Resonances. Therefore, based on the facts, I deduce that: *human brain is an intricate, miniaturized and vivacious projection of planet earth itself.* Any disturbance in the geophysical parameters is meant to be sensed by the human brain. Or to put it more simply, if anything happens to our beloved planet, no matter how subtle it is, our evolutionary instincts make us sense it.

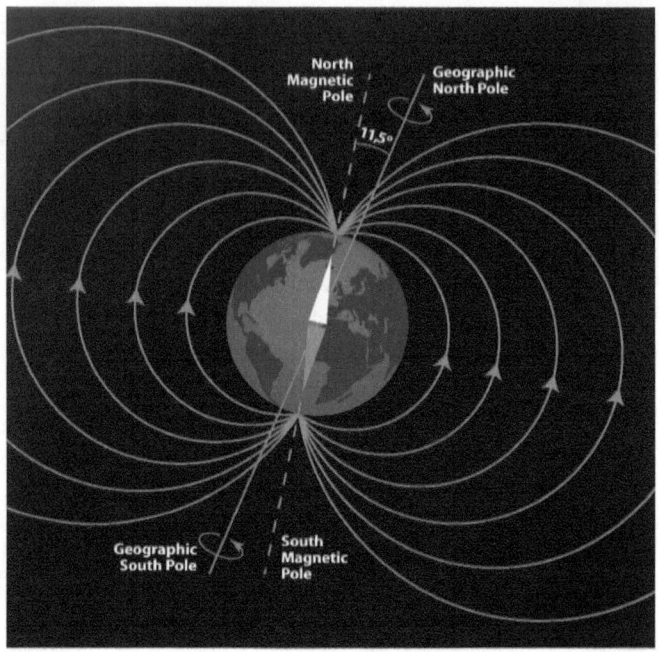

Figure 6.2 Earth's Geomagnetic Field

Every species is special and gifted in its own way. The earth-ionosphere cavity acts as the womb of mother earth, where she nourishes her children and sees them grow intellectually millennium after millennium. The only difference between mother earth and a human mother is that, after the umbilical cord is cut, the direct physical attachment is discontinued between the mother and her child, while on the other hand, human species is still too immature to break the earth-brain bondage.

Studies have shown that subjects living in isolation from geomagnetic rhythms over long period of time developed increasing irregularities and chaotic physiological symptoms, which were dramatically restored after the introduction of a very weak SR electromagnetic field. Early astronauts suffered behavioral disturbances until SR generators were installed in the spacecrafts.

A mother usually does not show more affection to one child and less to the other. But, on the contrary mother earth definitely loves humans more than any other species. The most significant evidence for this is the emergence of an inexplicable intellect in humans. And the geomagnetic field allows the humans to nourish this intellect, which is a product of millions of years of evolution.

The evolution of the physical, chemical and biological phenomena that have aggregated to produce complex organisms and the electromagnetic processes that emerged within their spatial boundaries have occurred while immersed in the geomagnetic field and the intensities and frequencies generated as Schumann resonances within the earth-ionosphere cavity.

The remarkable convergences between the temporal patterns and the intensities of the magnetic and electric field components for Schumann resonances found within both global human quantitative electroencephalographic activity and earth-ionosphere activity quite simply shows the significant existence of the earth-brain bondage.

Just like mirror neurons are the basis of the human capacity to empathize with another person, the brain's own electromagnetic characteristics enable the humans to empathize with our planet earth.

* * *

CHAPTER 7

GOD IS A FIGMENT OF YOUR IMAGINATION – EXPERIENCING GOD IN THE BRAIN

God is the most undeniably important, perpetual delusion of human evolution. The fact that the concept of God has survived so far in human minds while evolving with the human brain, implies its significance eloquently. Primitive men worshiped the mountains, the sun, thunder and as their cerebral neural network evolved, they started to worship more obvious humanlike forms of Gods, along with a few religious sectors of the society believing in the Supreme Energy source which has no form.

But this chapter is not at all about religion. It is neither on the argument about whether God exists or not. Rather, in this chapter I'll show the scientific basis of the concept of God, how it is hardwired within the human brain and how exactly it has been and still is necessary for most of the earthlings!!!

Well… to understand this extraordinarily long process we need to go back in time and look behind the curtain, how the Darwinian evolution paved the way for this mysterious concept, and more importantly how exactly some of the most important God-like experiences occurred in the human brain throughout mankind history! Experiences like Buddha's Nirvana, Moses's Ten Commandments, Muhammad's Revelation, Indian Sages' Vedas & Upanishad and many others have fueled the philosophical needs of mankind for ages.

Figure 7.1 Moses and The Ten Commandments

God is nothing but an essential figment of your imagination. All real and imaginary experiences are generated by the brain; experiences of all Sentient Beings, including God, are generated by cerebral activity. Mystics throughout history have claimed to experience visions and trans-like states, which they say come directly from God. But do they really come from God! In this chapter, I'll take you in a journey to explore the mysterious cerebral functioning behind religious/spiritual/mystical experiences (RSMEs).

Recorded history of neurological evidence shows that RSMEs are evoked by transient, electrical microseizures within the temporal lobes of human brain. Normally when we think of epileptic seizures, we think of someone convulsing and losing consciousness. But that's just the most known one type of seizure called grand mal seizure.

There's a whole other category of seizures, known as focal or partial seizures, that can cause a kaleidoscope of symptoms, such as the sense of oneness, complex hallucinations and feelings of fear, depression, and euphoria. Often, these seizures don't involve any convulsions at all. In some epileptics, a seizure can even invoke the presence of God.

The Russian Novelist, Fyodor Dostoyevsky, whose writings are among the world's greatest literature, had a rare form of temporal lobe epilepsy termed "Ecstatic Epilepsy". Dostoyevsky kept records of 102 epileptic seizures during his last two decades, which mainly occurred at night. Seizures which occurred in the daytime were often preceded by an ecstatic aura, which has led neurologists to theorize that he had temporal lobe epilepsy. Based on his experiences, he created characters with epilepsy in his four novels The Possessed, The Brothers Karamazov, The Insulted and Injured, and The Idiot. The Idiot is an example of how art can contribute to scientific observation.

Figure 7.2 Fyodor Mikhailovich Dostoyevsky

Dostoyevsky lets us see into the mind and emotion of the person with epilepsy through his character Prince Myshkin:

"He [Myshkin] remembered that during his epileptic fits, or rather immediately preceding them, he had always experienced a moment or two when his whole heart and mind, and body seemed to wake up to vigour and light; when he became filled with joy and hope, and all his anxieties seemed to be swept away forever; these

moments were but presentiments, as it were, of the one final second (it was never more than a second) in which the fit came upon him".

Dostoyevsky even recorded his own seizure experiences :

"For several instants I experience a happiness that is impossible in an ordinary state, and of which other people have no conception. I feel full harmony in myself and in the whole world, and the feeling is so strong and sweet that for a few seconds of such bliss one could give up ten years of life, perhaps all of life.

I felt that heaven descended to earth and swallowed me. I really attained god and was imbued with him. All of you healthy people don't even suspect what happiness is, that happiness that we epileptics experience for a second before an attack."

Likewise, historical data indicates that Joan of Arc, The Maid of Orleans from France, who is well known as the messenger of God was also an epileptic. She suffered from tuberculosis with a temporal lobe tuberculoma and tuberculous pericarditis. It is possible to explain Joan of Arc's experiences and behavior in terms of a widespread chronic tuberculous infection which became inactive in some organs' and calcified in others.

Calcification of the tuberculous mesenteric glands and chronic tuberculous pericarditis could account for the heart and parts of the intestines being intact after she was burnt at the stake. A tuberculoma (lesion) in the temporal lobe of her brain could account for her complex visual and auditory hallucinations of Archangel Michael, Saint Margaret and Saint Catherine.

Figure 7.3 Joan of Arc Being Burnt at the Stake

Until now there have been countless events of such hallucinatory God-encounters or experiences among the human population of the world. It was in the early 19th Century, Paris when it got first recognized by the scientific community that epilepsy is the root of all religious/spiritual/mystical experiences (RSMEs).

Every now and then we neuroscientists come across new speculations of mystical experiences. A specific conversion experience after an epileptic fit was reported in 1872. The patient believed that he was in Heaven. He would appear to have been depersonalized, as it took three days for his body to be reunited with his soul according to the patient. He expressed that he was now a new man, and had never before known what true peace was.

Most of the Temporal Lobe Epilepsy patients express the feeling of inexplicable bliss during the seizure. Many hear the voice of God. Everything starts to seem interconnected to them, and they finally understand the meaning of the universe. Many of them declare that God has given them different missions to complete, like building a temple or church or even reforming the whole world. They endorse strong personal beliefs in either specific cultural deities such as Allah, Brahma, Vishnu, Shiva, Christ, Virgin Mary or more exotic god surrogates such as "The Supreme Universe". But all these experiences are happening inside the brain, nowhere outside. It is just our mind playing tricks on us.

Just like a lesion in the temporal lobe, naturally occurring electromagnetic fields can cause RSMEs as well by impacting over the cerebral activity in the temporal lobe. For example, powerful meteor showers were occurring when Joseph Smith, founder of the Church of Latter Day Saints, was visited by the angel Moroni, and when Charles Taze Russell formed the Jehovah's Witnesses. The experience of God and all spirits can even be induced by artificial electromagnetic stimulation around the head.

Many neuroscientists have tried to find out the God-Spot in the brain that is responsible for RSMEs. But what we have found is that it is not just one, rather several major regions in the brain that are responsible for all kinds of God experiences.

Andrew Newberg, a neuroscientist at the University of Pennsylvania used single photon emission computed tomography (SPECT) to take pictures of the brain during religious activity. SPECT provides a picture of blood flow in the brain at a given moment, so more blood flow indicates more activity. One of Newberg's studies examined the brains of Tibetan Buddhist monks as they meditated. The monks indicated to Newberg that they were beginning to enter a meditative state by pulling on a piece of string. At that moment, Newberg injected radioactive dye via an intravenous line and imaged the brain. Newberg found increased activity in the frontal lobe, which deals with concentration; the monks obviously were concentrating on the activity.

But Newberg also found an immense decrease of activity in the parietal lobe. The parietal lobe, among other things, orients a person in a three-dimensional space. This lobe helps you look around to determine that you're 15 feet away from the window, 6 feet away from the door and so on. Newberg hypothesizes that the decreased activity in the brains of the meditating monks indicates that they lose their ability to differentiate where they end and something else begins. In other words, they become one with the universe, a state often described in a moment of transcendence or Samadhi.

Newberg found similar brain activity in the brains of praying nuns. So, it does not matter whom or what that religious activity is directed toward. Though the nuns were praying to God, rather than meditating like the monks, they showed increased activity in the frontal lobe as they began focusing their minds. There was also a decrease of activity in the parietal lobe, seemingly indicating that the nuns lost their sense of self in relation to the real world and were able to achieve communion with God.

The brain has two hemispheres. The left hemisphere is more of a logical and practical person, while the right hemisphere is the creative genius that creates and senses the supernatural. All your sense of self and the sense of the other are derived from the subtle but complex structural and neuroelectrical differences between the left and right hemispheres of the brain.

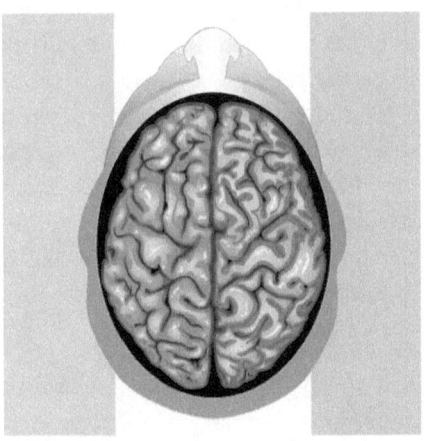

Figure 7.4 The Analytic Left and The Creative Right Hemisphere of The Brain

Whereas traditional left, more linguistic, hemispheric processes are strongly coupled to the sense of self, the transient intrusion of the right hemispheric equivalents are associated with the sensed presence of the supernatural, guiding angel or the supreme. So, basically you have two faces of your mind, but only one of them remains dominant in every person. Literally, half your brain is theist and the other half is atheist. So, what really happens when you die! Does half your brain go to heaven and the other half go to hell!!!

But jokes apart; human brain is hardwired to find meaning in everything it observes. That's why when you see an absurd painting, you can find your own meaning in it. A hundred people looking at a Van Gough would see the painting in a hundred different ways. It is inherent in human brain to make sense even out of nonsense. It is an evolutionary mechanism that developed through millions of years to keep the mind calm even in confusing times. When the brain lacks information on the scientific explanation for something, it tends to fill the gaps with supernatural explanation. It's nor stupid, neither foolish, rather it's really human to believe in the supernatural. Just like atomic energy has the potential to power a big city or to destroy one, God-experience has the potential to make better and confident human beings as well as to kill thousands of people in events like 9/11.

But does this mean that all the religious founders and the spiritual giants of human history were epileptic! The answer would be a straight no. Most of them had their own God-experiences through the natural means of prayer, devotion and meditation. On the other hand, temporal lobe microseizures triggered by brain lesion or planet earth's geomagnetic disturbances can evoke similar experiences as the religious founders.

However, whatever the cause may be, in all cases the deep spiritual experience takes place right within the complicated mesh of nerve cells of the human brain, without the intervention of any Almighty Being. And the content of the experience is formed by the brain based upon the person's own emotions, beliefs, intuitions, urges and conjectures. Thus the person's own distinct characteristics are imposed on the manifestation of the experience. Moreover, various neurochemical changes during the deep spiritual experience, leaves the person radically transformed. Often, the person comes out of the experience as a completely changed human being with certain unbreakable beliefs and intuitions about the universe. And all of this happens right within the neurons.

Now, one might wonder, what about the rest of the general human population who tend to be spiritual or religious by nature? Do all of them experience some sort deep spiritual encounter?

In majority of the general religious/spiritual population, what we see is more of a basic sense of religiosity/spirituality, than a deep spiritual conviction like the religious founders or the temporal lobe epileptics. Indeed they are religious/spiritual in their hearts. However, that is also not by choice, but driven by the evolutionary instincts. Darwinian natural selection has embedded the characteristic of religious belief in the brain circuits to survive the hardship of daily life.

Also, as humans spend the beginning of their lives under the guidance and care of parents, it is ingrained in their involuntary reflex to rely on a parental guidance all through the lifetime, either as a father figure like Jesus Christ or as a mother like Goddess Kali, or as a more abstract higher power such as Paramatman or Allah. Sometimes it's just good to know that someone or something is looking after you, guiding you in the right path and giving you the strength to survive the bad times.

* * *

BOOK II

Your Own Neuron

A Tour of Your Psychic Brain

INTRODUCTION

Neuroscience is still in infancy and so are humans as a species. We are just beginning to take baby steps on the journey of exploration through the beautiful and enigmatic maze of the human brain. The enigma within the human brain is way more exciting and thrilling than studying the collision of particles in the CERN super collider.

Being a neuroscientist, the synaptic interactions within the brain circuits thrill me more than the particle interactions of the physical world. That makes me think, if the laws of particle physics are the basic characteristics of the entire physical universe, then why can't the same properties be at play behind the cerebral interactions as well?

To be honest the human brain is not separate from the physical universe. Every molecular interaction and every synaptic jump of various neuro-modulators are actually way more entangled with the vast cosmos than we think. The brain itself is an amazing apparatus of space-time modulation. The physicists would say... *"in order to understand everything around us, it is necessary to see it through the glasses of a physicist"*. But I can perceive the inexplicable illustration of the entire cosmos within the human brain. Even so, what is so special about the 3 pounds lump of jelly that we carry around in our head!!!

I must be kidding right!!! Come on... we all know human brain is the dean of this entire institution of limitless human endeavors. It runs us all. Like literally all of us; the Picasso fan next-door, the physicist two blocks from your doorstep, the friendly microbiologist who keeps hitting on you, Steven Spielberg, my favorite Carey Grant, our dearest Erwin Schrodinger, Satyendranath Bose, Madam Curie, the father of psychoanalysis Sigmund Freud or your dearly beloved friend.

I'm not incorporating any other animals in the list other than the most domestic ones, i.e. humans, as for this book I need the best brain on planet earth. Perhaps you can already sense the spicy smell of modern physics in this preface.

I'm about to take you on a captivating journey to the paranormal corners of the human brain where dreams of modern physics are already a reality. I'm talking about the parapsychological phenomena such as Telepathy, Remote Viewing (often called as Clairvoyance), Mental Time Travel etc.

Just imagine how cool it would be, if all of this becomes possible without the use of any billion dollar particle accelerator. The only equipment that can already carry out such bizarre feats is the amazing cauliflower on top of you. Scientifically speaking, all these so called mystical and paranormal phenomena do not have a single bit of paranormal element in them. They already happen in your daily life, often without your conscious awareness or control. In this book, I'm going to show you that everything that happens in human life, no matter how bizarre it seems, there is always a scientific explanation behind it. Various phenomena seem to be paranormal as long as you don't know the scientific explanation behind them. Once you become aware of the scientific reality behind all your experiences you can eventually learn to utilize those experiences for a better outcome in daily life.

* * *

CHAPTER 1

YOUR VERY OWN
TELEPATHIC ABILITY

Have you heard of Quantum Entanglement in the physics community? Or perhaps you'd get it, if I ask you about Albert Einstein's spooky action at a distance. Now we are talking... right!!! Just imagine; two particles get so entangled with each other that they can share information, no matter the distance in between. Sounds cool!!! But why exactly I'm talking about Quantum Entanglement or in a broader aspect Quantum Physics? Well, that's because on our little planet and beyond, the game is all about particle interaction, whether it is inside the human brain or a non-living system.

But you know what! This sophisticated concept of the Quantum Physics world actually has an amazing influence over the human brain. Quantum Entanglement is the magical event that makes your subconscious often sense what your dearly beloved one is feeling thousands of miles away from you. No matter how far away you are through space-time, you are biologically designed to sense the person's feelings and thoughts, with whom you are emotionally attached too much. It could be your daughter, son, father or in most cases your lover and mother. This is what's widely known as Telepathy.

But, before we continue, I must clarify something. Such experience has nothing to do with the supernatural world. It is all very biological.

Basically, Telepathy is an amazing method of communication through which separate neural networks of different brains talk to each other irrespective of the distance in between. It happens to everybody once in a while. There is nothing supernatural or paranormal about it.

Every now and then you experience something like this. How often does it happen that you are thinking about phoning your friend and quite surprisingly that specific friend rings you before you have even dialed the number!!! Although in my last book I explored this exact matter of mind-to-mind connectivity, in this chapter I push the boundaries a little further into the inexplicable journey through the enigmatic characteristics of the amazing human apparatus of communication, i.e. the brain.

Quantum Entanglement between emotionally attached brains becomes possible by the amazing gift of Mother Nature, what I have termed "Earth-Brain Bondage". All of our smartphones today are connected to a dense network of information throughout the entire planet. Our beloved planet earth is filled with electromagnetic fields which are the medium of information exchange between our smartphones. Now imagine your brain as a smartphone and earth's geomagnetic field as the medium of information exchange. In order to send a selfie to your friend, you need to have his or her contact details. Likewise to exchange information between brains you need to have the access code to the other brain. And here the access code is nothing but your deep emotional attachment to the other person.

And now answer a simple question. What happens to your phone signal when there's a terrible thunderstorm outside? It just becomes really unstable. This exact phenomenon happens to your telepathic ability whenever the geomagnetic field is disturbed due to solar storm, cosmic rays or increased lunar gravity pull. That's why most of the unconscious telepathic information exchange between two brains occurs during geomagnetically calm days.

Figure 1.1 The Sun is constantly emitting high energy particles; earth's geomagnetic field shields us from the harmful solar radiation so that we can live safe and sound on this planet

Talking about geomagnetic field, there are so many geophysical factors that impact over your cool sci-fi ability to communicate with your dear one's mind. The geomagnetic field of planet earth acts as an invisible force that protects us from the sun's harmful radiation.

Although the biggest responsibility of earth's geomagnetic field is to shield us from the solar wind, it has way more inexplicable influence over the living systems of our planet, especially over the human brain.

Since the beginning of human species, it has been and continues to be immersed within the geomagnetic field. Its general intensity averages around 50,000 nT (0.5 gauss). Disturbances in the geomagnetic field lead to physiological problems in humans. These disturbances occur primarily from increased solar activity (Coronal Mass Ejection, CME) mediated by changes in the velocity and density of particles in the interplanetary magnetic field, increased lunar gravity interaction on full moon and man-made ionospheric alterations. These fluctuations take place for some minutes and their magnitude of intensity ranges in the order of 25 to 1000 nT. And whenever the magnitude of disturbance intensity exceeds 25 nT, it impacts over the human biology and psychology way more eloquently than one can imagine. These effects include alterations in occurrences of migraine headaches, strokes, glaucoma pains, joint swelling (objectively measured), blood clotting, skin conductivity, tissue permeability, embolism risks, thyroid activity, heart attacks, schizophrenia, paranoia, psychotic breaks, epileptic seizures etc.

Psychotic breaks during various lunar phases are very common, especially in the full moon. This mysterious phenomenon gave birth to a terrifying mythological character in the 15th century Europe, known as werewolf.

Several studies have evidently shown that a full moon really can bring out the beast in humans, turning them into wild savage animals. The so called Dr Jekyll within each one of us can literally transform into Mr. Hyde in a full moon. This mythological concept has a genuine psychiatric condition at its foundation, which according to DSM is a cultural manifestation of schizophrenia. Clinically this condition is known as Lycanthropy. Here the person's own belief and cultural perspective play a crucial role in his or her delusional state of mind where one reports being in the process of transforming into an animal.

The impact of geomagnetic disturbance over the brain during full moon triggers intense psychotic breaks in mentally unstable patients and sometimes make them something like bloodthirsty animals. A study conducted in Australia showed that 91 emergency patients with violent, acute outbursts comparable to werewolves were admitted to a hospital north of Sydney. And a quarter of these outbursts occurred on the night of a full moon, double the number for other lunar phases. Some of these patients attacked the staff like animals, biting, spitting and scratching. The patients had to be sedated and physically restrained to protect themselves and everybody around them.

Planet earth is way more entangled with every molecular interaction within the human body than you can imagine in your wildest dreams. Therefore any disturbance to our planet's geophysical parameters can show signs of physiological and psychological stress in human body.

But negative impacts of the earth-brain bondage are nothing compared to the limitless benefits of it. One of those amazing benefits is your very own telepathic ability. Yes, you heard right… Mother Earth has already given all humans the ability to communicate with each other's mind. All humans are real psychics, but not the kind you know of.

Basically, one way or another we all human beings are psychological freaks. Come on… let's be honest!!! How many times have you just thought of your dearly beloved, and suddenly you happened to receive a text from that special person! Or may be, you just woke up with a dream about a family member and found out that the person had been in some sort of crisis. How many times it happens to you that you are about to say something to your spouse, but he/she just spits out the exact same words right before you!!! Do you really think that it is all mere coincidence! This is just the craftsmanship of nature that is encoded within the molecular map of the human brain. Every synaptic interaction within the neural network of the human brain carries the signature of our planet earth.

The geomagnetic field works as the most crucial medium of connectivity between two emotionally attached human brains. Human brains are subconsciously connected to the planet earth, while on the contrary all animals are quite consciously aware of every single change in the geophysical state. As I have elaborated in my last book The Art of Neuroscience in Everything :

"Human brain is the most complicated mystery of our planet. The womb of mother earth is filled with mysterious mechanisms. These mechanisms make the creatures of this planet uniquely special in one way or another. Every living organism on earth is always connected to the geophysical state of the planet. We see the beauty of this correlation everywhere around us, most apparently in animals. For example, an accurate navigation system has been gifted to various earthly creatures by our planet, which is called "Magnetoreception". It works in pace with the earth's geomagnetic field and helps the migratory birds to find their way back home.

To explain a little more we can say that the avian magnetic compass is a complex biological mechanism with many surprising properties. The basis for the magnetic sensing is located in the eye of the bird, and furthermore, it is light-dependent, i.e., a bird can only sense the magnetic field if certain wavelengths of light are available. Specifically, many studies have shown that birds can only orient if blue light is present.

Likewise, animal behavior prior to earthquakes is quite captivating to human imagination. All animals instinctively respond to escape from predators and to preserve their lives. A wide variety of vertebrates already expresses "early warning" behaviors right before an earthquake. So it's plausible that a seismic-escape response could have evolved from this already-existing genetic predisposal. An instinctive response following a Primary wave seconds before a larger Secondary wave is not a huge leap, so to speak, but what about other precursors that may occur days or weeks before an earthquake! In fact there are precursors to a significant earthquake that we have yet to learn about, such as ground tilting, groundwater changes, electrical or magnetic field variations. Indeed it's possible that some animals could sense these signals and connect the cognitive phenomenon with an impending earthquake."

During the exchange of information between two cerebro-magnetic fields of two human brains, the geomagnetic field of planet earth makes way for quantum entanglement between the two intricate quantum systems, i.e. the two brains.

Scientifically, entanglement has been described as excess correlation between separated parts of a quantum system that may exceed the boundaries of light across space and time. In case of telepathically connected brains, the separate brain quantum systems along with the medium of geomagnetic field, give rise to a giant universal quantum system. Here all the human brains and planet earth's geomagnetic field together act relentlessly as one global quantum system until it gets disrupted by the disturbance in the geophysical parameters especially in the geomagnetic field lines.

The cerebro-magnetic field of a human brain consists of local photon emission. Here, photons, the electromagnetic wonders that display both particle and wave properties, appear to be the key elements for

Quantum Entanglement between two non-local brains. Through the medium of the earth's magnetic field two separate fields of biophoton emission from two human brains can share information at vast distances.

Does that mean all human brains can share information with each other without the use of any online cloud service! Technically yes, but that may become the basic biological characteristic of the future human beings. For now, it is more likely for two emotionally attached or biologically related brains to exchange thoughts and emotional stimulations because more regions of their neural networks become synchronized as the intimacy grows.

Figure 1.2 The Global Brain with billions of living brains as its neurons

Just imagine the entire planet earth as a ginormous brain with its own characteristic quantitative EEG and all the living brains as the neurons of that global brain. Now you would think that there are so many different brains of different species!!! In this case, brains of various species build different types of neural networks, where the network containing the human brains are the most advanced one and others are just comparatively dim except for their several outstanding abilities (magnetoreception, earthquake predictability etc).

Now let's look into the most advanced neural network of the wonderful global brain. If you look at the biological design of a neuron or nerve cell, it possesses a cell body (soma), many dendrites and an axon.

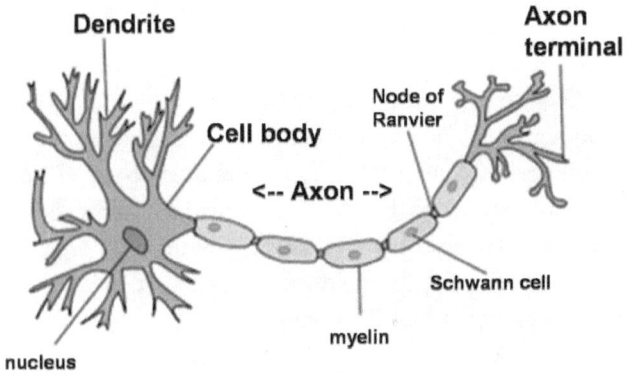

Figure 1.3 A Neuron or Nerve Cell

And the connecting regions between two nerve cells [could also be a nerve cell and another cell (neural or others), but it is irrelevant to this chapter] are called synapses. With the use of neurotransmitters a synapse permits a neuron to share electrochemical information with another neuron.

Now we'll explore the global brain parameters that are characteristically similar to the human brain. In the gigantic characteristic design of the global brain, the geomagnetic field acts as the synaptic region of communication between two human brains just like two neurons in the brain itself. Then where do you find the neurotransmitters through which the exchange of information is done!!! Well... in this global brain it has every single ingredient that permits one brain to share information with another. Here, the planet's very own Schumann Resonance behaves as the neurotransmitter. Schumann Resonance is the nutrient with which Mother Earth has been nourishing the entire human species from the very beginning.

EEG pattern of the human brain waves are the most eloquent evidence of planet earth's Schumann touch over cerebral activity. Lewis B. Hainsworth of Western Australia seems to be the first researcher to recognize the relationship of brain wave frequencies with the naturally circulating rhythmic signals of the planet, known as Schumann Resonance (SR).

In 1952, a German physicist, Dr. W.O. Schumann, suggested that the space between the surface of the earth and the ionosphere acts as a resonant cavity, somewhat like the chamber in a musical instrument. Energy for the SR is provided by lightning. Lightning pumps energy into the earth-ionosphere cavity and causes it to resonate at frequencies in the ELF range. In the normal mode descriptions of Schumann resonance, the fundamental mode is a standing wave in the Earth–ionosphere cavity with a wavelength equal to the circumference of the Earth.

In 1957, Dr. Schumann calculated the SR frequencies and fixed the most predominant standing wave at about 7.83 Hz. The frequency values of the SR signals are determined by the effective dimensions of the cavity between the Earth and ionosphere. Thus, any event that changes these dimensions, also changes the resonant frequencies. This means too much fluctuation in the Schumann Resonance due to geomagnetic disturbances disrupts the information exchange between two human brains in the network of the global brain. While on the other hand, calm geomagnetic state of planet earth enables appropriate synaptic information exchange within the global brain.

So, you can see that the Global Brain is a giant Quantum System, in which all the human brains behave like separate particles, which simultaneously emit characteristic information from each thought and emotional stimuli. Just like in Bose-Einstein Condensate all the particles are in a single quantum state (i.e. the ground state), together all the human brains act as a "condensate" which forms a massive Global Brain quantum state.

According to Bohr (1958), the simultaneous emission of two particles with opposite spin from an atom produces a condition such that altering the spin of one instantaneously reverses the spin of the other no matter what the distance. Entanglement is associated with non-locality that has been described by Cramer (1997) as enforced correlation between separated parts of a quantum system that are outside of the boundaries of light velocity across space and time in order to ensure the parts of the system maintain equilibrium. It might even be considered as a trans-temporospatial application of Newton's third law *"for every action (or force) there is an equal and opposite reaction (or force)."*

Two brains with histories of mental or physical proximity, might become entangled by processes as quantifiable and experimentally reproducible as those displayed by pairs of particles. Which means two brains that are psychologically or physiologically related, eventually generate an entangled quantum state, in which a single cortical activity of one brain has the potential to influence cortical activity in the other brain.

Many of us neuroscientists take the amazing phenomenon of telepathy seriously. In fact these mysterious wonders of human brain fascinate us. Venkatasubramanian and his team were the first to report real-time functional magnetic resonance imaging (fMRI) of the changes within the cerebrum during procedures that could be defined as "telepathy". Although different people perceive the term "telepathy" differently, from a neuroscientific stand-point it has a really simple definition. We define it as information acquired from a distance through mechanisms of photon field information exchange or, from a more modern neuroquantum perspective, as "distant intentionality".

Now you must be wondering about the brain regions that might be responsible for your extraordinary telepathic ability. Venkatasubramanian and his team found that the greatest activation within the brain of Mr. Gerard Senehi (a renowned mentalist) while he was thinking about the image that was being drawn by another person in a different room occurred in his right parahippocampal gyrus. Here you should notice, that Senehi was thinking about the picture while it was being drawn by another person, which means, the data his cerebral photonic field received were not just directly from the picture itself, rather photon emission from the other person's brain was also involved. Hence, you can see a telepathic connection between these two brains, which conveyed electromagnetic data about the image that was being drawn. The parahippocampal gyrus is a major multimodal integrator of sensory information from the association areas of the neocortex and is intimately involved with visuospatial processing. And the right one, to be more specific is responsible for integration of all kinds of psi information, like telepathy, clairvoyance, precognition and premonition.

Which means that whatever telepathic data the cerebral photonic field receives through Quantum Entanglement, are forwarded to the right parahippocampal gyrus for further integration.

Then the right parahippocampal gyrus makes you aware of the information either consciously in a vibrant form or subconsciously in several figments of confusing thoughts. But unless you are a true Mentalist like Mr. Gerald Senehi, those telepathic visions or thoughts won't occur to you as vividly as they are shown on TV.

Mostly it comes to you as a random figment of thought or emotional sensation. You know what I'm talking about, right!!!

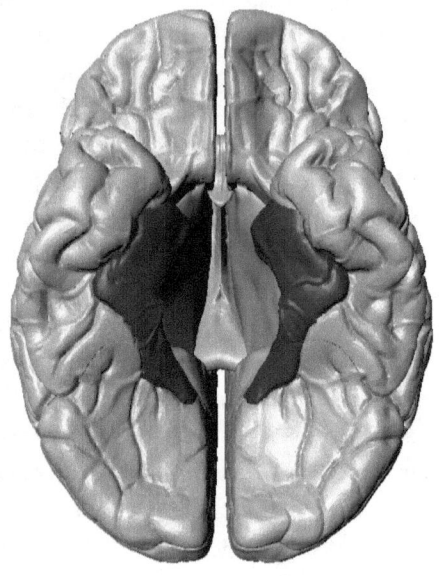

Figure 1.4 The Parahippocampal Gyrus in both hemispheres are shown as darker areas; the right one is our man for telepathy

For example, sometimes quite out of the blue, you feel so blue. You keep asking yourself, what's happening to you, since you cannot find any physical or psychological cause of your sudden mood swing. In this kind of situation you must ask yourself, is it really yours, or someone else' emotional state, what you are sensing. Think again… could it be your romantic partner or your sibling or any other person with whom you have strong emotional attachment, who might be upset!!! The psychological cue from another person far away can also trigger physiological response in your body. So, quite romantically speaking, if you truly love someone, that someone's illness can literally become yours. Likewise, in a conversation with a dear one, that person often grabs the word right from your mouth or your thoughts and vice-versa.

The beauty of these extraordinary abilities is that you cannot force them to occur, rather it all happens quite naturally. The best you can do to utilize these abilities in your life is that you grow more attachment with the one you care about the most. Get close to each other and let your own brains do the magic.

* * *

Chapter 2
Mental Time Travel

A few months ago a person, very close to my heart had an accident. After she recovered from the devastating trauma, she mentioned something really intriguing. During the accident, she was in the car with her uncle, while her mother and sister were several miles away from her at home. Just before the time of the accident, her mother suddenly became terribly anxious and felt as if something really bad was about to happen. Likewise, her sister felt a choking sensation in her throat during the time of the accident. But none of them had any idea about what was really happening to them, until they got the bad news that their beloved one had been in an accident.

All of us humans experience this kind of paranormal phenomenon once in a while, but most people are just shy to express, as they fear that they might be considered as crazy. But aren't we all more or less crazy in our own way!!! Well... I admit that I am. And to be honest, you cannot be perfectly sane and a scientist, both at the same time.

Anyways, let's get back to the topic. From chapter 1, you get the perception on how physical and psychological closeness can build a telepathic network between human brains. This telepathic network allows a mother to sense the imminent threat to her daughter and a sister to feel the crisis of her sibling. Now you might ask, what about her father? Well... her father didn't feel anything significant. That's because of her more obvious physiological connection with her mother and sister, as she and her sister were once part of their mother's body. But notice that the experiences that the mother and the sister went through, were characteristically different. The mother experienced the prevailing danger even before it actually happened, as if she travelled mentally several minutes into the future. This is what we scientifically call Premonition, a subcategory of Mental Time Travel. While on the contrary, the sister showed physical symptoms during the time of the accident, which was the product of Quantum Entanglement between sibling cerebral activities.

In this chapter you are about to embark on a journey of Mental Time Travel. Yes... you heard right. I'm talking about Time Travel, but not anything like HG Wells' novel The Time Machine. To consciously or subconsciously travel through time with the mind's eye, humans don't really need a separate Time Machine. The only machine you need in order to become a time traveler is born with you. It is your own brain. We call this phenomenon as Mental Time Travel, because your body remains where it is, while your mind travels through the fourth dimension, i.e. time, and brings you information in the form of visions or thoughts or some kind of emotional alertness. Most of the times, this happens to people without any conscious choice. It just happens.

Your brain itself is a living time machine. Its intricate neural network is the enigmatic circuitry of the time machine. Some traditional neuroscientists think that Extra Sensory Perceptions (ESP) like telepathy, premonition, precognition and clairvoyance violate the laws of physics. According to them, precognition and premonition contradict the principle of causality, as an effect does not happen before its cause. But come on guys, let's not be so cynically orthodox. Even the laws of physics are not persistent themselves. They keep changing. Just like in a recently published paper, our very own Big Bang Theory has been shown to be wrong after all. Physicists Ali and Das at the University of Lethbridge in Alberta, Canada, have shown in their paper published in Physics Letters B that the Big Bang singularity can be resolved by their new model in which the universe has no beginning and no end. Their new model implies that the universe may have existed forever.

Memorizing facts and then regurgitating them into carefully crafted words is not science people. It's intellectual bulimia. Real science happens when we explore what we don't know. The first law of understanding the human brain and the mind within, is to be an explorer. As I mentioned earlier, neuroscience is still in infancy. And with every single experiment, we are discovering new aspects of the human brain and its unbound capabilities and wonders. Studies have already shown that the human brain is pretty scientifically engineered throughout the period of evolution to be in touch with the fellow brains at all times. And when your dearly beloved is in danger your brain lets you know that your beloved one is in some kind of crisis.

But the phrase Mental Time Travel is also used for another crucial cognitive process, which is called Episodic Future Thinking. It is a concept introduced by Atance and O'Neill in a 2001 article. Essentially, episodic future thinking involves using the mechanisms and resources of episodic and semantic memory to project one's self into future situations, that is, to simulate how future events involving the self would be organized and unfold based on the knowledge one has in episodic and semantic memory. Thus, imagining the future is constrained by experiences in the past. So, it is pretty much human and frankly speaking, not so cool as Premonitive and Precognitive Mental Time Travel.

We can say that Episodic Future Thinking is a basic human ability, while Premonition and Precognition are supposedly superhuman characteristics. You are like the very famous Doctor Who, who can move through space-time without any boundaries riding the Tardis of mind.

These mysterious characteristics of the human brain are perceived as "psi phenomena". The term "psi" denotes anomalous and enigmatic processes of information transfer that are currently not so well explained with known physical or biological mechanisms. In this book I'm illustrating those mysterious cerebral psi processes in a vivid scientific manner. Psi has two distinct and vivid variants, which are Premonition and Precognition. But you must be thinking, what's the difference really between Premonition and Precognition!!! It's all so freaking same psychic stuff, right!!! Not really.

Although, both deals with future awareness, they are characteristically different. As defined by the Merriam-Webster dictionary, precognition is *"clairvoyance relating to an event or state not yet experienced"*. We'll explore Clairvoyance vividly in the chapter 4. Derived from the Latin term 'praecognoscere', precognition can be literally translated as 'to foreknow'. Typically associated with dreaming, this term refers to the act of seeing an event before it occurs. Though a conscious act, it is not necessarily controlled or self-initiated. The majority of the time precognition is reported to occur in dream states, during which the recipient clearly has no control over the vision. Other times precognition may be induced while awake by trance or other means of deep meditation. It can come as both straightforward and absurd visions. Which means, dreaming or not dreaming, precognition is more of a vision and image oriented experience.

But remember, the world of dreams itself is an enigmatic one with its own interpretation. Most of the dreams are either a mental simulation of a virtual reality where your most intense desires come true, or just a projection of the events of your daily life. A rare portion of your dream might actually be precognitive. So don't be abruptly presumptuous in considering all dream contents as precognition.

On the other hand, premonitions are associated exclusively with the emotional realm. Typical premonitions occur in relationship to negative events to come. One may feel an overwhelming sensation of anxiety or foreboding before a tragedy. The definition for premonition can again be traced back to Latin roots. Originating as praemonere, premonition is described by Merriam-Webster as a *"forewarning"*, or *"anticipation of an event without conscious reason"*, or *"to warn in advance"*.

The emotions felt during premonition are always directly related to an event to come, and mirror the emotions that will be felt as it takes place. Premonitions are never controlled or self-induced, rather they occur spontaneously without conscious effort.

We neuroscientists have been running studies and experiments on the psi phenomena for a long time. In the recorded literature of psi experiences, between the years 1920 – 1967, 57 telepathic cases and 56 precognitive cases were reported.

The first comprehensive study on psi phenomena was published in a book called "Phantasms of The Living". It was published in the year 1886, undertaken by the Society for Psychical Research, London. The book consisted 702 cases of telepathy, precognition, premonition, hallucinations, clairvoyance etc. The study was done by Edmund Gurney, Frederic William Henry Myers and Frank Podmore.

Spontaneous subjective experiences of premonition and precognition involving death, sickness or crisis of friends and relatives are frequently so intense that the experients (individuals who experience) record them in diaries or tell friends or family members. When premonition and precognition experiences occur, they usually involve extreme alterations in autonomic tone, such as feelings and visions of impending doom. When these experiences occur at night or during sleep, they usually cluster around dream periods. If they occur just before dream periods, in the most typical cases the experient suddenly wakes up. These patterns have not changed since the time of the first published work in the field, Phantasms of The Living.

Our approach to these intriguing experiences is that they are evoked by environmental stimuli. These stimuli are correlated with death, sickness or crisis of a beloved one. Here your beloved one doesn't actually transmit the information orienting the crisis, because in case of premonition and precognition that person might not even be aware of the impending crises, rather you, the percipient or experient get influenced by the correlated environmental stimuli.

The natural psi phenomena demand an environmental stimulus that can penetrate buildings and propagate for thousands of kilometers. The environmental stimulus also have the capacity to directly influence the brain of the percipient. This environmental stimulus carries all of the information contained in the experience.

How the environmental stimulus, i.e. the geomagnetic field of planet earth carries the psi information even at vast distances has been vividly illustrated in the first chapter.

So, let's look into something deeper and more fascinating. In case of precognition and premonition, the information is literally being carried through the fourth dimension, i.e. time.

What could be the possible explanation for this kind of time travel!!! I can only hypothesize two plausible ways, in which this kind of time travel could take place.

First one is that, human brain is not just entangled with another emotionally or biologically related brain at only vast distances in three dimensional space, rather it remains entangled in the fourth dimension as well to some extent. Now, you might wonder, if the emotionally or biologically related brains or to be more specific, cerebro-photonic fields are entangled in four-dimensional space-time, then how come we do not receive info about the near future events of a beloved one's life all the time!!! That's because the sharing of information in time is triggered only by the impact of a dreadful event over the future brain of your dear one.

The second plausible means of mental time travel is the environmental stimulus. In this case the geomagnetic field is capable of carrying the information of the dreadful event from a few moments in the future, when the event is actually taking place and bringing it to your cerebral proximity at present. As a result, this incoming distress signal from the future evades your brain like a virus and evokes potential discharge in the neural network of various regions of the brain. We'll explore those regions in a while.

Premonition and precognition usually affect individuals in an aversive way. More than 75% of all psi cases are anxious in nature. The experient often reports physiological symptoms as well as psychological effects of apprehension or anxiety. They include fluctuations in body temperature, trembling, sweating and alterations in heart rate.

In various studies, about half of the experients have reported that they were asleep when they had the experiences of precognition or premonition concerning death, crisis or illness of family members or friends. Of those who were asleep more than half reported, that they were dreaming or had just finished dreaming. Another half reported that they suddenly woke up. Psi experiences that occurred when the experients were not sleeping, were associated with states of decreased vigilance. They include dozing or automatic tasks like washing the dishes, cleaning the floor, and walking without thinking.

Among all the psi cases of scientifically recorded literature, more than 74% of the putative agents were actually dying. By the way, did I tell you about the putative agents? I didn't right!!! Okay, here is the explanation for that. I mean, the explanation to putative agents, not to my forgetfulness.

Just like, a 'percipient' or 'experient' is the person who's having the psi experience, putative agent is the person who's the central theme of the psi experience.

The pattern of correlated actual dying condition of the putative agents during psi experiences is evident for both contemporary, historical and cross-cultural samples.

Let's hear some of the people who have experienced the same kind of psi phenomenon that you sometimes do. These reports were published in a scientific magazine.

Report 1

"Nothing seemed unusual about the evening of January 5, 1969. I was visiting in my kitchen in Coos Bay, Ore., with my sister-in-law, Mrs. Bud Arnold, and her friend, Ruby Kirbs. Hot coffee steaming in cups before us. Had no feeling of impending disaster, no sense of inner disturbance, until suddenly my face flushed and I felt ill.

I excused myself and went into the bathroom to run cold water over my wrists and bathe my hot face. Feeling better soon, I returned to the table – but in a matter of minutes I was out cold. My sister-in-law told me later they carried me to the daveno and placed a cold cloth on my forehead. Then she said I opened my eyes and simply stared into the space. I don't remember this – or anything else, for that matter. I must have remained this way for 15 minutes, we later decided, and when I came to the time was exactly 10:30.

Next I was deathly ill. I vomited many times that night, between snatched of restless sleep.

Each time I awoke I felt like crying and several times I did. I kept thinking of my mother, Sharan Shaw, who lived in Chemult, Ore., 200 miles away. For no same reason, thoughts of her brought a terrible depression.

The next morning at seven o'clock Mother called me. She had been crying and was extremely upset. She told me that my sister Barbara had died the night before in Spokane, Wash. – at exactly 10:30 – Coos Bay, Ore."

Report 2

"One Saturday in the fall of 1955 a strong feeling came over me that I should visit my friend Jane Bertelman at once, instead of the following evening as we had planned. We lived only a few blocks apart in Washington, D.C., but it was raining and I had little inclination to go out.

But the feeling nagged at me. I'll phone her, I told my unseen taskmaster. The peremptory answer came, 'Don't phone! Go. Go NOW!'

Reluctantly I dressed for the inclement weather – sweater raincoat, umbrella – and at last I was outside. Now actually on my way I no longer felt the urgency. Six blocks to walk in a pelting rain! What had gotten into me anyway? My steps slowed and I sheepishly thought of turning back. However, my tormenting prompter still was on the job. 'Hurry, Hurry!' So I hurried.

Fortunately the way was all downhill. Taking advantage of every step-saving shortcut I soon was pushing open Janes's small iron gate. Ordinarily the gate's creak would have set her two dogs into a wild clamor. Now all was silent, even after I twisted the old-fashioned doorbell.

'What a wild goose chase,' I muttered. 'She isn't home'.

'Rap!' came the command. I rapped and then removed my glove to knock with my ring against the glass. Still no answer.

'Again!' I knocked once more. This time the curtain was plucked aside. White of face and haggard-eyed Jane was trying to turn the lock.

Oh, I'm sick, she gasped, at last opening the door.

One sniff was enough.

It's gas! I cried and leaving the door wide open I held my breath and rushed to the water heater in the basement to turn off the gas cock I knew to be defective.

Thanks to some unseen force, I had made it in time."

The relationship between the experient and the putative agent is a major factor in psi experiences. And, just for the record, pardon my really tasteless technical words like experient and putative agent. Sometimes we scientists get really cocky with the usage of technical terms.

Anyways, talking about relationship, most psi cases involve members of the immediate family, such as husbands, wives, mothers and fathers etc. The second most frequent class involves members of the distant family (aunts, uncles, grandparents) and close friends. A small number of cases involve peripheral friends or acquaintances and strangers. I'm not done yet. There's more. A frequently forgotten or ignored class of psi involves death or crises of the experient's favorite pet.

The distance is not really a matter here. Studies have shown psi phenomenon occurring at a distance of up to several hundred kilometers.

The brain region that is responsible for precognition and premonition, is the parahippocampal gyrus. But here the temporal lobes of the brain play a crucial role. Like you read in the last chapter, the parahippocampal gyrus is a major multimodal integrator of sensory information from the association areas of the cortex and is intimately involved with visuospatial processing.

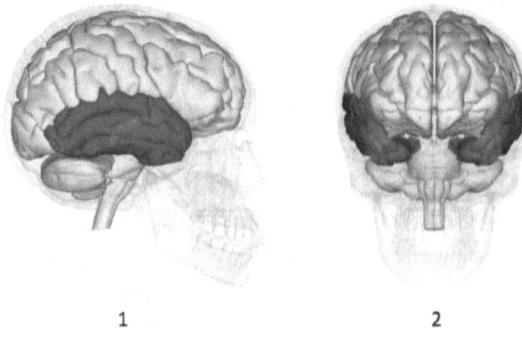

1   2

Figure 2.1 Temporal Lobe (1. Lateral view, 2. Front view)

Then you might ask what do the temporal lobes do really? Well, the temporal lobes are your brain's special effect expert. That means, they make the experience of precognition and premonition mystical enough to arouse your both physical and mental alertness.

Mystical experiences are the domain of the temporal lobes. Especially, the right temporal lobe of your brain is capable of creating all kinds of mystical experiences, when stimulated either naturally by a perturbation of the geomagnetic field in the surrounding environment or by a tiny brain lesion pushing against the temporal region. For example, in many medically recorded cases, patients have reported that they hear their dead parents, brother, sister or even God's voice. What really happening here is that, their own wish and instinct are getting exaggerated by the stimulation of the temporal lobe.

Sometimes this mysterious apparatus of the brain generates premonition or precognition like phenomenal experiences without any actual psi data from the parahippocampal gyrus. But when the temporal lobe is stimulated due to parahippocampal modulation of the psi info, then your entire physiological and psychological state of being starts responding to that info. Then you suddenly become emotionally anxious and alert with the sensation that something really freaky is about to happen. That's because the temporal lobe contains structures that either directly or indirectly have a major influence over the physiological conditions, that in turn effect the emotional state.

Like other cortices of the brain, the temporal cortex is organized into functional columns measuring about 250 to 500 micrometers in width. Two mutually connected major structures exist beneath the cortex. They are the hippocampus and the amygdala. The hippocampus is primarily associated with memory (consolidation of information from short-term memory to long-term memory), spatial navigation and occurrence of dreams or altered states. Dysfunctions of the hippocampus can seriously alter memory and release dreams into waking states. The amygdala is most well known for its association with emotional experiences. Stimulation of this structure produces intense feeling of meaningfulness, usually of cosmic or religious significance.

Which means when the amygdala gets stimulated either naturally or artificially, everything starts to make sense. As if you suddenly understand the meaning of the entire universe and everything in it. It really is an awesome sensation.

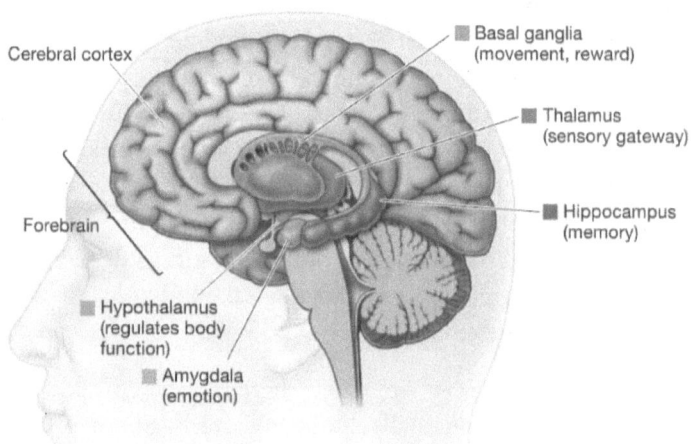

Figure 2.2 The Limbic System

When the hippocampus is also stimulated along with the amygdala, in this state, you amazingly find yourself being absolutely one with your surrounding objects. It's like everything around you, your favorite chair, your beloved tablet, this book, the street in front of your house, the neighbor's dog and all other things, no matter how insignificant they are, start to feel like one entity with you. In the ancient Sanskrit scriptures from India, this state of being is highly celebrated. The ancient sages called this state "Samadhi". It's the ultimate of all trance states.

In addition, the amygdala, because of its connections to the hypothalamus, has strong influence over the autonomic nervous system. It is the visceral seat of the emotions. As a result, during psi experiences or any other mystical experience the temporal lobe influences over the autonomic nervous system through the amygdala and hypothalamus.

Direct surgical stimulation of the subcortical temporal lobe structures, especially hippocampus and amygdala can generate identical experiences, many of which constitute the essential features of psi experiences. When small focal currents are induced within these tissues, the percipient may hear meaningful messages, have visions, report alterations in space and time (déjà vu), or even feel as if the self has left the body(out-of-body experiences). These experiences are often products of an epileptic brain, but they may also be elicited in a healthy normal brain.

Actual premonition and precognition experiences would involve natural stimulation of the temporal lobe by the stimuli that are related to the environment. And the psi information that is forwarded to the temporal lobe is modulated by the parahippocampal gyrus.

Precognition and premonition are Mother Nature's most enigmatic gifts to mankind. But, like it goes with all other gifts that humans get their hands on before they are mature enough to handle them, we are yet to learn how to distinguish and utilize these extraordinary abilities. It's a long shot, but it's totally worth it.

* * *

CHAPTER 3
TWINS & TELEPATHY

Every now and then you may hear that twins can pick up each other's thoughts. They can experience simultaneous pain, joy and all other emotions. In various scientific studies, it has been shown that identical (monozygotic) twins are more prone to such telepathic connectivity than non-identical or fraternal (dizygotic) twins.

Twins can feel each other's pain both physical and psychological even while being thousands of miles away from each other. Research on exceptional experiences of connectedness in twins, such as telepathic experiences, dates back to the work of Sir Francis Galton who was knighted in 1909.

Since then, many claims have been made that some twins who are physically remote from each other show a sensitivity to each other's thoughts, emotions, actions, pains, and sensations. The authors of major survey in the United Kingdom found that 39% of twins believed they might have *"the ability to know what was happening to their partner,"* and a further 15% were convinced of it. Identical twins were twice as likely as non-identical twins to report these experiences. These findings were given support by a Swedish survey in which 60% of twins reported telepathic experiences. Furthermore, frequencies were significantly greater among identical twins and significantly related to the degree of attachment between the pairs of twins.

Twins are often able to communicate through telepathy more easily than other ordinary siblings or unrelated people. Some of the earliest research on twin telepathy was done by a twin, Professor Horatio Newman, head of the zoology department at the University of Chicago. He had what he considered telepathic experiences with his twin brother and published a book called Twins and Super-Twins in 1942. The section on telepathy discussed identical twins who were mystified by the way they could communicate with each other without any verbal exchange.

In 1961, Robert Sommer, Humphry Osmond, and Lucille Pancyr interviewed fourteen pairs of twins and seven single members of a twin pair to see how many of them reported experiences of telepathy. Twelve out of the thirty-five participants believed that they could communicate telepathically with their twin. They made statements such as,

"We both think the same things at the same time,"

"I can tell what her feelings are,"

and, *"When my twin goes out, I can imagine what he is doing and see the place, like right now, even if I've never been there or seen the place described."*

Telepathy happens frequently between closely connected twins during crisis. The term crisis telepathy was coined after several dramatic accounts such as the following:

Martha Burke felt as if she *"had been cut in two"* one day in 1977 when a searing pain crossed her chest and abdomen. Hours later she discovered that her twin sister had died in a plane crash halfway across the world.

Similarly, in July 1975, Nita Hurst's left leg became agonizingly painful as bruises spread spontaneously up the left side of her body. She later discovered that her twin, Nettie Porter, had been in a car crash at the very same time four hundred miles away.

As I mentioned earlier, monozygotic or identical twins are more likely to have telepathic connectivity of minds and bodies than dizygotic or non-identical twins. So, let's explore a little bit about the biological background of the twins, in order to find out the scientific basis behind the intense Quantum Entanglement between identical twins.

Monozygotic or identical twins occur when a single egg is fertilized to form one zygote (hence, "monozygotic") which then divides into two separate embryos.

Dizygotic or non-identical twins (also known as "fraternal twins", "dissimilar twins", "biovular twins", and, informally in the case of females, "sororal twins") usually occur when two eggs, fertilized by two different sperms are implanted in the uterus wall at the same time. The two eggs, or ova, form two zygotes, hence the terms dizygotic and biovular emerge. Fraternal twins are, essentially, two ordinary siblings who happen to be born at the same time, since they arise from two separate eggs fertilized by two separate sperm, just like ordinary siblings.

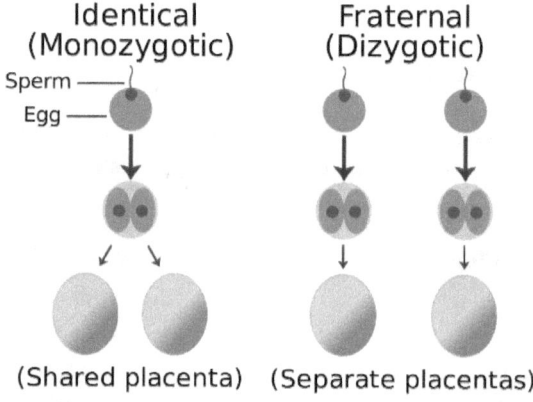

Figure 3.1 Monozygotic and Dizygotic Embryo Formation

So, you can see that unlike dizygotic twins, monozygotic twins come from a single egg. This gives the advantage of lifelong Quantum Entanglement between the two particle-like embryos. At a molecular level, monozygotic embryos maintain a functional connection throughout their whole lifetime irrespective of the distance between them.

This Quantum Connectivity allows the identical twins to feel each other at great distances and even communicate with each other without any smartphone. In non-identical twins and other siblings the Quantum Entanglement is mildly present but not as intense as in identical twins.

In case of identical twins, their genetic kinship and similarity, or more specifically the fact that they come from the same egg, plays a crucial role in building up a profound telepathic connection between them. While on the other hand, in non-identical twins, other ordinary siblings, spouses, very close friends, their emotional attachment with each other brings their brain wave EEG in sync, during events of crisis. However, brain wave synchronization is much more extreme in identical twins.

The early reference to telepathic connectivity between twins comes from the work of psychologist Sir Francis Galton in the 1870s. But again, it was not until the publication of Phantasms of The Living in the year 1886, when the world of mysterious psi phenomena got the attention of the scientific community. This book was a massive treasure chest of various case histories which were written in illustrative detail and utmost accuracy.

Recently I watched an episode from the sci-fi TV series The Flash, Season 1. In this specific episode after getting exposed to particle accelerator explosion two individuals get immersed into one person. In this person, the body remains the same as the body of the younger individual, while the consciousness of the older and more experienced individual turns out to be the dominating consciousness of the newly formed personality. So, the consciousness of the younger individual sinks down at the bottom of the subconscious mind of the newly formed human being and the molecular structure of the older individual's body somehow gets merged within the matrix of the younger body.

Now you must be thinking, why the hell am I just talking about all these sci-fi jargon!!! Well, that's because the episode illustrates the wonderful Quantum Entanglement between two persons with all of its minuscule details, that is usually seen in identical twins.

The fun begins when our friends at Star Labs attempt to separate those two individuals. And off course, since they are one single entity even at a molecular level, so it's not easy to do that with a bone saw. We need something much more sci-fi-ish.

After a lot of twists and turns, Cisco, Caitlin and Wells come up with a cool looking prototype of a Quantum Splicer. With this amazing device they successfully separate the young and hot Ronnie Raymond and the elderly physicist Dr. Martin Stein in perfectly healthy state. But after the separation they start showing signs of Quantum Entanglement. For starters, totally out of the blue, Dr. Martin Stein develops a craving for pizza, which happens to be Ronnie's favorite. They start feeling each other's pain, literally. They even can tell each other's exact location in time of distress. When the army abducts Dr. Stein by sedating him, Ronnie while being many miles away from him starts to feel dizzy. In the secret military base, Dr. Stein is now getting electrocuted with a taser and miles away from him, Ronnie is feeling the taser shock in the Star Labs with no actual taser attack.

Then, in order to know Stein's location Ronnie curves the word "WHERE" on his arm with a sharp knife and quite excitingly the wound in the shape of "WHERE" shows up on Stein's arm. Not so surprised, Stein recognizes what's happening to his arm. He then takes a look around the room for any kind of sign. And he finds out the label "Military Research Facility # 27" on the wall. So, he uses his intellect as well just like Ronnie and gently starts hitting his hand cuff against the table in a specific pattern of dots and dashes. Ronnie on the side suddenly starts feeling the taps on his arm. In a blink of an eye Wells recognizes them as Morse Code. Cisco deciphers it and finds out the number 27. Ultimately Barry Allen a.k.a The Flash rescues Dr Stein from the army base. Except for the sci-fi mumbo jumbo of fusing two human beings into one, then separating them, and the end part of sharing info by cutting one's self, the story really elaborates on the mysterious characteristics of Quantum Entanglement between two persons.

Quantum Entanglement has already been described neurologically in the first chapter. So, you already know how exactly the wonderful phenomenon works in order to form the connectivity of minds, especially between two biologically related or emotionally attached minds.

Connectivity of minds between identical twins is the ultimate form of Quantum Entanglement in human beings. And even if you do not have a twin brother or sister, you can still feel the mental state of your siblings who might be thousands of miles away if you are mentally attached enough. It's a gift to mankind from Mother Nature, so you might as well nourish it.

* * *

CHAPTER 4

CLAIRVOYANCE OR REMOTE VIEWING - A FUTURE WITHOUT SECRETS

Imagine a future without any kind of secret. A world free from secrets and confidentiality. Sounds like fantasy, right!!! Actually not at all. Building that future already started long ago, since the time when we humans cultivated the ginormous cauliflower in our head. A future where you will be able to visualize anything and everything that is happening around the globe. Yes, you heard right.

Evolutionarily you already have that ability to access information from another person's brain. In telepathy, you get to look inside another person's brain, like you read in the previous chapters, as if you can look inside the mind of another person and get to know even his or her deepest and darkest secrets.

But this chapter is about another extraordinary ability of the human brain. We call it "clairvoyance" or "remote viewing". This awesome ability of your brain shows its mettle when you sometimes just quite subconsciously access visual imageries of remote events or objects. Or to speak simply, in clairvoyance you have the ability to perceive events that might be happening thousands of miles away from you. That's why it's called "remote viewing", which means to view remote events.

Traditionally the difference between clairvoyance and telepathy is that, the later involves thoughts of another person, while the former involves objects and events. The environment is filled with information. The procedure of "remote viewing" within the brain ostensibly incorporates those information from the environment. And you already have absorbed the perception from the previous chapters, that the human brain is always immersed in this vast geomagnetic field of information. If only we could see that marvelous world of information around us, how cool would it be!!! But you know what, you don't need to see that information with those two eye balls in order to access the information. Your mind can sense it even when you are not at all aware of it.

In many traditions people consider this kind of ability as the work of the "Third Eye". But remember, this third eye stuff is total poppycock. There is no physical existence of any "third eye" between your eyebrows, which miraculously opens up every now and then and makes you see things.

Speaking of third eye, a nice little story just popped up in my mind. The most amazing illustration of the third eye comes from the ancient Indian mythological texts "Matsya-Purana" and "Shiva-Purana". Heaven was in distress by the wrath of a demon called "Taraka". He couldn't be defeated except by Lord Shiva's Son. As a final resort, the Indian version of Cupid, Kamadeva was assigned to break Shiva's meditation and make him engage in coitus with wife Parvati. Kamadeva, accompanied by his wife Rati sneaked into Shiva's abode in the Himalayas.

While lord Shiva was engrossed in his meditation, Kamadeva made repeated attempts to arouse passion in the heart of lord Shiva, but his actions were no avail. Right then, Kamadeva saw Parvati arriving accompanied by her friends. She was looking divine in her beauty. Just at that moment lord Shiva too had come out of his meditational trance. Kamadeva thought that it was the most appropriate moment to have a go.

Kamadeva struck lord Shiva with five of his 'Kamabana' (arrows of love) which did have a deep impact on him. Lord Shiva was struck by the awesome beauty of Parvati and his heart became full of passion for her.

But at the same time he was surprised at the sudden change in his behavior. Quite obviously, his prefrontal cortex kicked in, i.e. if he had any. He realized that it was an act of that stupid pervert Kamadeva.

Lord Shiva looked all around him. He saw Kamadeva standing towards his left side, with a bow and arrows in his hands. Now he was fully convinced that it was indeed an act of Kamadeva.

Shiva was boiling with rage. Kamadeva became terrified, he started remembering other deities, but before the deities could come at his rescue the third eye of lord Shiva got opened and something like high intensity laser beam or flames of fire came out of it. Kamadeva was incinerated to ashes in seconds. Sounds just like the laser beam that comes out of the two primary eyes of Clark Kent and Cyclops from X-Men.

Figure 4.1 A Cambodian Shiva with the Third Eye

Rati, the wife of Kamadeva wept in grief. The deities arrived and consoled her by saying that by the grace of lord Shiva, her husband would be alive once again. After that the deities went to lord Shiva and did his worship. They told him that it was not the fault of Kamadeva, as he had acted in accordance with the aspirations of the deities. They also told him the mystery of Taraka's death. The deities then requested him to make Kamadeva alive once again.

Lord Shiva told the deities that Kamadeva would take birth as the son of Krishna (you know.... The Indian Casanova) and Rukmini in the era of dwapar. Until then Kamadeva would live in a disembodied form, which is why he is also known as 'Ananga' (an- = without; anga = body, "bodiless"), or 'Atanu' (a- = without; tan = body). Later he is reincarnated in the womb of Krishna's wife Rukmini as Pradyumna.

Talking about ancient Indian texts, you must have heard of the mystical concept of 'Chakras', a.k.a. 'Kundalini Chakras'. Third eye is one of those chakras. It refers to the Ajna Chakra. But, come on guys, even a toddler can tell you that there is no freaking third eye that magically appears every now and then between your eyebrows.

The tiny dramatic story of Shiva tells us about the misperception concerning the so called third eye. We can say, that the phrase "third eye" is just a symbolic representation of your mind's eye, which by the way doesn't have the ability to emit high intensity laser beam.

Many investigators from various fields of the society believe that humans had in far ancient times an actual third eye in the back of the head with a physical and spiritual function. Over time, as humans evolved, this eye atrophied and sunk into what today is known as the pineal gland.

Really..............!!!

You must be kidding..........!!!

So far we don't have any proper scientific evidence to back up this ridiculous theory.

In the far ancient times, humans had tail, which over time had mostly vanished and what remains of it is coccyx, or tailbone. We basically all are monkeys.

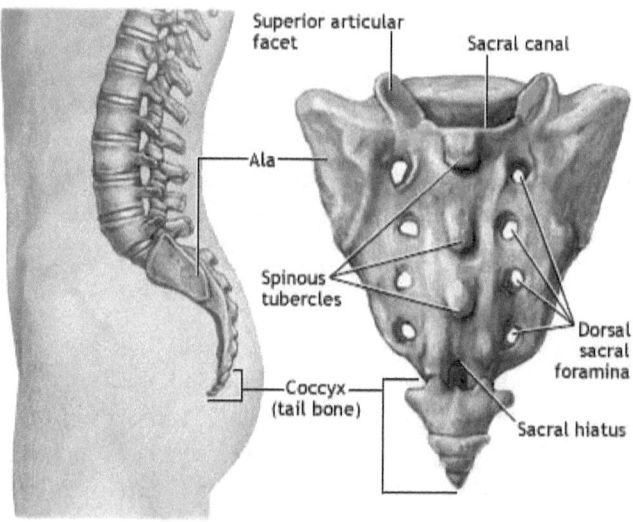

Figure 4.2 The Coccyx or Tailbone

Other than the tailbone, there are several leftover parts in the human body. These structures are called vestigial structures. And pineal gland, which is often mistook as

the third eye is considered to be one of them. Let's look
into some of these evolutionary leftovers.

Our ancestors needed strong molars for mashing up
and chewing plant material. This is why many of us
often develop wisdom teeth, also known as third
molars. Theoretically, they could still be used for
chewing, but in most people they cause pain and
infection.

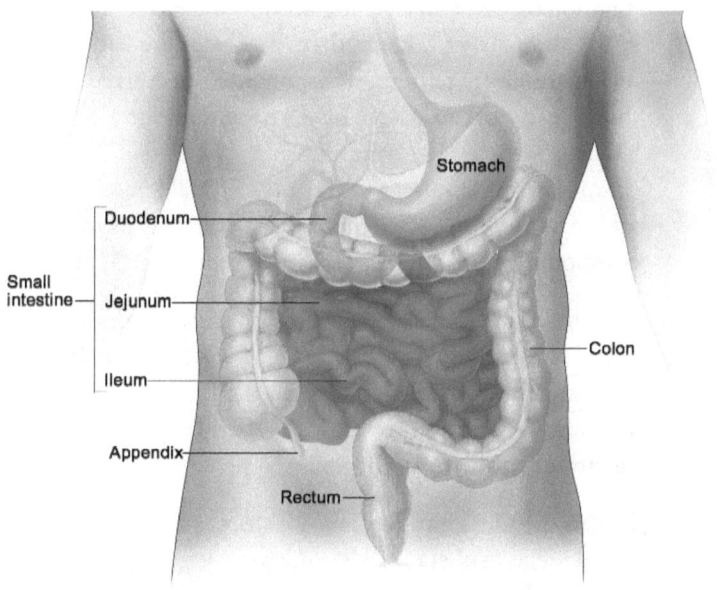

Figure 4.3 The Abdomen Region showing the Appendix

Another leftover from our plant eating ancestors is the
vermiform appendix, which is an organ attached to the
large intestine. A similar sac is much bigger in other
animals than it is in humans and is used to aid in

digesting high cellulose diets. When the appendix gets inflamed in humans, it causes severe pain in the abdomen. This is why these vestigial structures are almost always removed when they begin to come in.

And as for pineal gland, it is definitely an interesting little endocrine gland which is not a useless vestigial structure. It happens to be light sensitive, but doesn't have the ability to make humans see real things that might actually be happening around the globe. However it is the major gland that produces melatonin to modulate sleep pattern. It is wrongly perceived as the third eye because it also produces an amazing hallucinogenic compound called DMT (Dimethyltryptamine). DMT induces hallucinations that are very intense and vivid. This wonderful psychedelic compound is produced every night during REM sleep.

Post-traumatic stress or near death experiences involve excessive production of DMT, which helps in healing from the trauma. This counts for the bizarre paranormal imageries during near death experiences. It will be illustrated in the next chapter.

It can produce a profound effect unlike any other psychedelic substance, seemingly sending the user into **another dimension** and allowing the user to converse with higher dimensional entities or experience a full perception of reality. This is why, pineal gland is perceived as the third eye or even as the seat of consciousness. But remember that, everything you see under the influence of DMT, is merely hallucination. So it is nothing like the visions of the mind's eye, like precognition, premonition and clairvoyance.

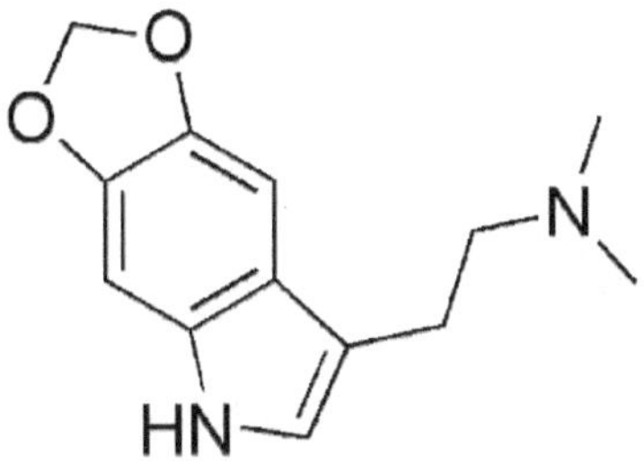

Figure 4.4 Dimethyltryptamine or DMT

The biological two eyeballs capture images that are actually in front of them, but the field of vision of your mind transcends all physical limitations in the three dimensional space and sometimes (in case of premonition and precognition) in the fourth dimension as well. The influence of the pineal gland over this ability is close to nothing.

Human perception is derived from the intracerebral processing of multiple, parallel sequences of electromagnetic (photonic) information after transduction of physical stimuli by sensors. In case of clairvoyance, electromagnetic information in the geomagnetic field acts as an environmental stimulus feeding info of a remote event into the brain.

Remote viewing is exactly like Quantum Entanglement between two brains, except, here we have only the percipient, and no living putative agent on the other side. So, what does the percipient brain get entanglement with? Quite frankly, this is where it gets even more exciting. The entire world around us is filled with photonic information. And your brain itself has its own photonic field. The most fundamental characteristic of the biophotonic field of human brain is that it can share photonic information with the surrounding electromagnetic field, in this case the geomagnetic field.

Cognitive perception is correlated with cerebro-magnetic or photonic field information. In case of remote viewing or clairvoyance, the percipient cerebro-photonic field becomes susceptive to the geomagnetic field information of a remote event.

As you have seen in the first chapter, the right parahippocamapal region of the cortices is responsible for integration of all kinds of telepathic as well as clairvoyant information. Especially the parahippocampal gyrus is a visuospatial integrator. So, the visual clairvoyant imageries are first processed by the right parahppocampal gyrus, then forwarded to the temporal lobe for further integration. During the clairvoyant experience the influence of the temporal lobe over the autonomic functions of your body through the limbic system makes you feel like you are actually observing the remote event while being physically present.

So far, many scientific studies have been done on individuals to record cerebral activity while they were experiencing distant information of a remote event. Basically, the electromagnetic field of your brain gives you the access to the environmental information, which consists of photonic data of events all around the globe. And some brains are more sensitive to the exchange of environmental info than others.

To clarify something, there are very few scientifically proven true psychics all over the planet. I'm just using the term 'psychic', because it is the most understandable even by the layman. While these rare people can consciously access psi information from the surrounding geomagnetic field, you often quite out of the blue become subconsciously aware of specific events that might be happening literally anywhere in the world. With this kind of characteristics of the human brain, there is simply nothing in the world that can remain secret.

The idea that human brains access components within the whole of the field generated by the billions of human brains immersed within the geomagnetic field is similar to the Vedic concept of the Akashic Record. However, these psychic abilities of the human brain do not require any spiritual assumption or paranormal element, rather they are quite intrinsic features of the Mother Nature.

* * *

Chapter 5

Near Death Experiences - Visit to The Spirit World

O thou the last fulfilment of life,
Death, my death, come and whisper to me!

Day after day I have kept watch for thee;
for thee have I borne the joys and pangs of life.

All that I am, that I have, that I hope and all my love
have ever flowed towards thee in depth of secrecy.

One final glance from thine eyes
and my life will be ever thine own.

Rabindranath Tagore

Death has long fascinated poets, artists and philosophers around the world. The mystery of death has always remained one of the most unfathomable concepts in all human cultures. Many consider it as the gateway to a higher dimension. The most rudimentary notion is that, it leads the soul to a realm, where there is only peace and happiness. This kind of perception eases the intense tension of death. Many people even anticipate death with the hope of getting rid of all the sufferings and pain. From this anticipation, rise various vivid extraordinary experiences, when a person survives a life-threatening trauma. These experiences are called Near Death Experiences (NDE).

NDE consists of a number of special elements such as out-of-body experience, pleasant feelings, seeing a tunnel of light, deceased relatives, angels, butterflies or a life review. These experiences mostly occur during traumatic events such as cardiac arrest (clinical death), shock after loss of blood, coma following traumatic brain injury or intra-cerebral hemorrhage, near-drowning, or asphyxia.

In one study of those who had a near death experience, 50% reported an awareness of being dead, 24% said that they had an out-of-body experience, 31% remembered moving through a tunnel, and 32% reported meeting with deceased people. Moreover, while it is common anecdote that NDEs are associated with feelings of euphoria and bliss, only 56% associated the experience with such positive emotions, and some even reported negative experiences.

Many of us so-called intelligent human beings grow up with the absurdly false idea that death represents a new beginning, a passing to an afterlife, where we are reunited with our loved ones and live eternally in a utopian paradise. This is common across most theological doctrines. NDEs only make these doctrines more deep-rooted in the mind of a person who encounters them without the awareness of the real neurological phenomenon acting behind them. Such bizarre experiences can be found across cultures in literature dating back to ancient Greece. Remember, there is real physiological basis for all kinds of near death experiences. You don't require a spiritual or religious component to explain these experiences.

So, let's go sightseeing the paranormal world of the human mind where NDEs are born.

One of the most frequently reported features of near death experiences is an awareness of being dead. This specific feeling is not however limited to NDEs. For example, an intriguing and rare neuropsychiatric condition that also creates the delusional awareness of being dead, is 'Cotard' or 'walking corpse' syndrome. This syndrome was named after the French neurologist Jules Cotard. He described the condition as *"le délire de negation"* or *"the delirium of negation"*. It results in a delusional reality that one is either dead or immortal.

In 1880 Jules Cotard reported the case of a 43 year old lady, Mademoiselle X who believed that she had *"no brain, nerves, chest or entrails and was just skin and bone"*, that *"neither God nor the devil existed"* and that *"she was eternal and would live forever"*. The syndrome is described to have various degrees of severity, ranging from mild to severe. In a mild state, feelings of despair and self-loathing occur, whereas in the severe state the person with Cotard syndrome actually starts to deny the very existence of the self.

In 2007 McKay and Cipolotti published a report on a 24 year old patient called LU. LU repeatedly thought that she was in heaven, even though she was actually in National Hospital, Queen Square, London and that she might have died from flu. The delusions diminished over a few days and were gone after a week.

It sounds like zombies, doesn't it!!! Well… yes, this is the realest thing you can get closely to the concept of zombies. Cotard syndrome is a rare disorder in which nihilistic delusions concerning one's own body are the central feature.

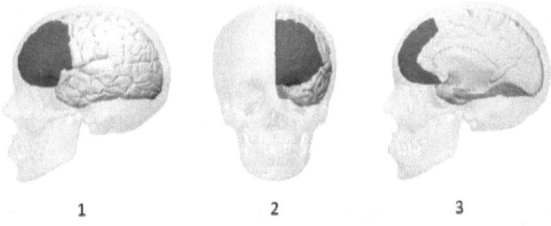

Figure 5.1 Prefrontal Cortex (1. lateral view, 2. front view, 3. medial view)

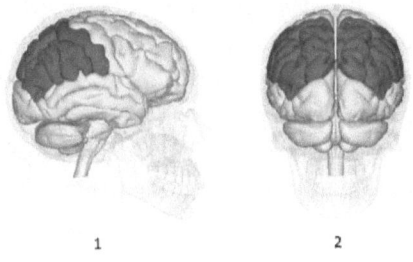

Figure 5.2 Parietal lobe (1. lateral view, 2. posterior view)

The medical literature indicates that the occurrence of Cotard delusion is associated with lesions in the parietal lobe as well as the prefrontal cortex. It is not listed as a specific disorder in DSM-V, as it is typically viewed as a symptom of other underlying disorders.

Cotard syndrome can be treated with various antipsychotic, antidepressant and mood stabilizing drugs along with electroconvulsive therapy (ECT) and psychotherapy. So, let me ask you a question. When a person is in this kind of delusional state that he or she is already dead and doesn't need either treatment or food, should we respect that delusion and let that person actually die out of starvation, just like Mademoiselle X died in the year 1880?

I guess, you already got the answer from your inner voice.

So, as you can see, in walking corpse syndrome a person actually walks around and carries out daily activities just like a normal healthy person, but with the delusion of being dead. This delusion goes away with treatment. While on the contrary, the awareness of being dead in NDE occurs when the body, especially the brain is traumatized. After the body revives from the traumatic episode, many people claim that they were dead during the episode. And in most cases, the person pleasures the memories of that experience rest of his or her life.

Another crucial phenomenon of NDEs is moving through a tunnel of light. And you know what…. just hearing about the marvelous tunnel of light, many people tend to jump to the conclusion that the light is coming directly from heaven. But, in fact, experiencing the tunnel of light can be explained by visual activity during retinal ischemia, which occurs when the blood and oxygen supply to the eye is depleted. This experience has widely been studied on fighter pilots.

Dr. James Whinnery, a chemistry professor with West Texas A&M and a specialist in aerospace medicine during his 30 years in the service, was charged with answering a question of practical concern to the Air Force: Why were so many pilots blacking out in the cockpits of high-powered F-15 and F-16 jets?

It had been the Air Force's policy to ground any pilot who suffered loss of consciousness while flying. But Whinnery and his fellow researchers discovered that perfectly healthy pilots were blacking out. It turned out the rapid acceleration of the jets was to blame.

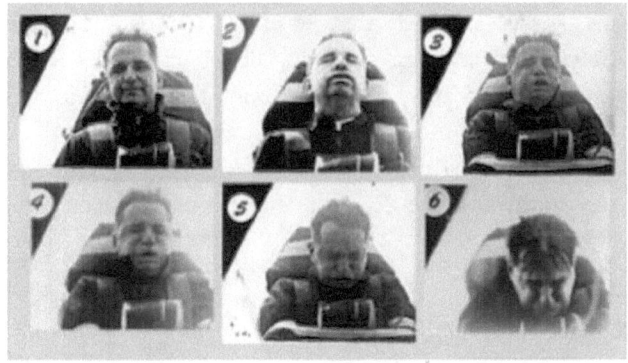

Figure 5.3 Impact of g-force on a fighter pilot

He undertook a long-term study of this phenomenon, which became known as Gravitation Induced Loss of Consciousness (G-LOC). He discovered that a healthy pilot could function normally when exposed to an average of 2.5 - 3 g-forces. But the pilots of fighter jets were regularly being exposed to up to 9 g's.

Whinnery and his colleagues placed pilots in gravitational centrifuges and examined how the G-forces affected their brains and bodies.

Figure 5.4 Gravitational centrifuge to simulate g-force

The study showed that during periods of rapid acceleration, a pilot's blood flows into his lower extremities and away from his brain. This results in lack of oxygen in the brain and leads to a period of unconsciousness that lasts on average 12 seconds, followed by another 12 seconds of disorientation.

During that twenty four seconds the pilot doesn't have control over the aircraft. Twenty four seconds was average. Some people were out a lot longer. The pilots were asked to describe the symptoms they experienced while in the centrifuge or in flight. A clear pattern emerged. The first thing they experienced was tunnel

vision, which results from a lack of blood flow to the
eyes and is a precursor to a blackout. Then, while
unconscious, many of the subjects experienced short,
vivid dreams, or dreamlets. The dreams were often
about past experiences or family and friends, just like a
life review.

Figure 5.5 Tunnel vision

In addition to dreamlets, a significant number of pilots
reported having out-of-body experiences. The subjects
described the sensation of floating above their planes
and looking down at their bodies. Sometimes the
dreamlets and out-of-body experiences were
accompanied by feelings of euphoria and warmth.

Whinnery's research led the Air Force to reconsider its
policies about pilots suffering blackouts. Now, losing
consciousness is not usually grounds for
disqualification. He also contributed to the safety of

fighter pilots by helping develop redesigned cockpits and anti-g suits.

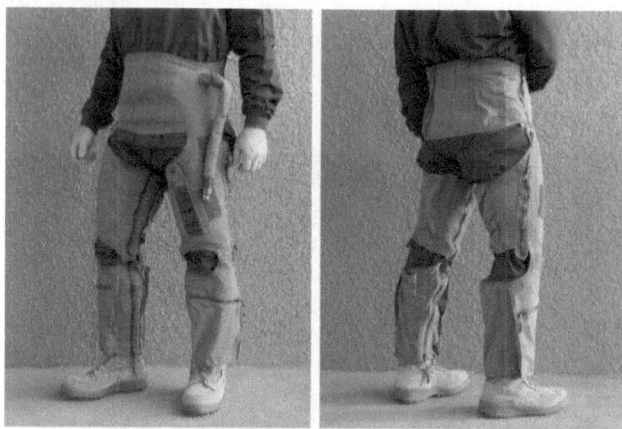

Figure 5.6 Anti-g Suit

An anti-g suit or simply g-suit is a special garment and generally takes the form of tightly-fitting trousers, which fit either under or over (depending on the design) the flight suit worn by the aviator or astronaut. The trousers are fitted with inflatable bladders which, when pressurized through a g-sensitive valve in the aircraft or spacecraft, press firmly on the abdomen and legs, thus restricting the draining of blood away from the brain during periods of high acceleration. The anti-g suit adds 1 more g to the pilot's g-tolerance limit.

Tunnel vision has been associated with hypoxia, i.e. oxygen loss. Whinnery's experiment on fighter pilots also provides ample support for other near-death experience research.

The most philosophically significant feature of NDEs is the out-of-body experience. During an out-of-body experience, the person seems to have the ability to perceive, feel emotions, create new memories of a higher dimension, but their "self" is not within their body. Rather, they leave the body and view it from above, which is called "own body perception". In other words, their center of experience is located above their physical body. Out-of-body experience consists of amazing vivid cognitive elements. These cognitive perceptions often have transcendental features like visiting another realm or world with strange creatures or angels. By the layman such higher realm can easily be perceived as heaven, because of its seemingly realistic qualities. And, quite unfortunately sometimes even experts from various fields of the scientific community get deluded by these hallucinations and perceive them as real. For example, a recent near-death experience of a neurosurgeon inspired him to write a book about his captivating trip to heaven. So, you can imagine, how real such an experience feels, that even a neurosurgeon seems to believe it as real!

Perhaps, sometimes even the smart people get tricked by the mind and act stupid. Whether heaven exists or not, that is irrelevant to this book, but the realm that a person perceives during a traumatic episode is just the brain's own way of keeping the person at ease by simulating a virtual environment of serenity and well-being.

A famous example is that of Pam Reynolds (Sabom 1998). She suffered a brain aneurysm that required a surgery that forced the blood in her head to be drained, not unlike oil from a car. The doctors taped her eyes shut and molded headphones to her ears while playing clicking sounds at greater than 100dB in order measure her brainwaves. Her body temperature was then lowered to approximately 19 degrees Celsius and the brain surgery commenced once her brainwaves were measured at zero and her heart was stopped. During the surgery, (she said later) she had the sensation of *"popping out of her head"*.

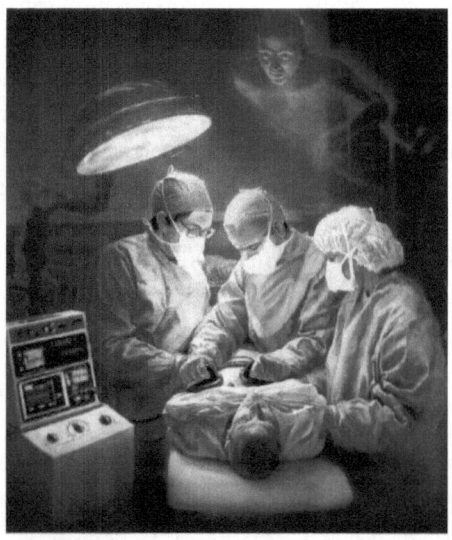

Figure 5.7 Something like this happens during out-of-body experience

She explained she was then viewing her body from above while floating from the ceiling. She could hear the conversation between the doctors discussing blood vessels in her legs. She even remembered the music that the doctors were playing during her surgery. She then felt her body being transitioned down a dark tunnel towards a pinpoint of light while hearing her grandmother's voice calling for her. After hearing her voice, all fear subsided and she drifted down this tunnel willfully. Her uncle also accompanied her during this experience. Once some time had passed, she journeyed back down the tunnel and into her body and had no recollection after that.

A year later, assuming it was a hallucination, she spoke to the surgeon about the experience. He showed her the video and audio recording of the surgery. Amazingly, they found that the conversation she recited actually happened during the surgery and the song that she remembered, "Hotel California" by the Eagles, also was playing in the room as they put her body in a cardiac standstill.

None of these experiences requires a single piece of spiritual or religious element. Because none of these is real. It's all in the head. Just like you saw earlier that tunnel vision is caused by lack of oxygen. Out-of-body experience, perception of a higher dimension and encountering strange creatures, angels and diseased relatives have their roots deep within the intricate physical framework of the human brain.

The magical own body perception in out-of-body experience is caused by the impaired functioning of the temporoparietal region, thereafter rise vivid cognitive visit to another realm and encounters of angels, unearthly creatures and dead relatives. The encounters are sometimes more religious in nature, as people claim to have perceived many mystical beings, spiritual giants of human history and even God. But again, these experiences are not real either, they are the work of an amazing psychedelic compound called Dimethyltryptamine (DMT). DMT is produced in the pineal gland.

But before we visit the misperceived heaven, let's look into the formation of own body perception in out-of-body experience (OBE). An OBE is defined by the presence of three phenomenological characteristics, which are: disembodiment (location of the self outside one's body), the impression of seeing the world from a distant and elevated visuo-spatial perspective (extracorporeal egocentric perspective) and the impression of seeing one's own body (autoscopy or own body perception) from this elevated perspective. How the brain creates OBE is freaking awesome.

OBEs not only are found in clinical populations but also appear in approximately 10% of the healthy population. Let's hear a subjective experience of a healthy person.

"I was in bed and about to fall asleep when I had the distinct impression that "I" was at the ceiling level looking down at my body in the bed. I was very startled and frightened; immediately [afterward] I felt that, I was consciously back in the bed again."

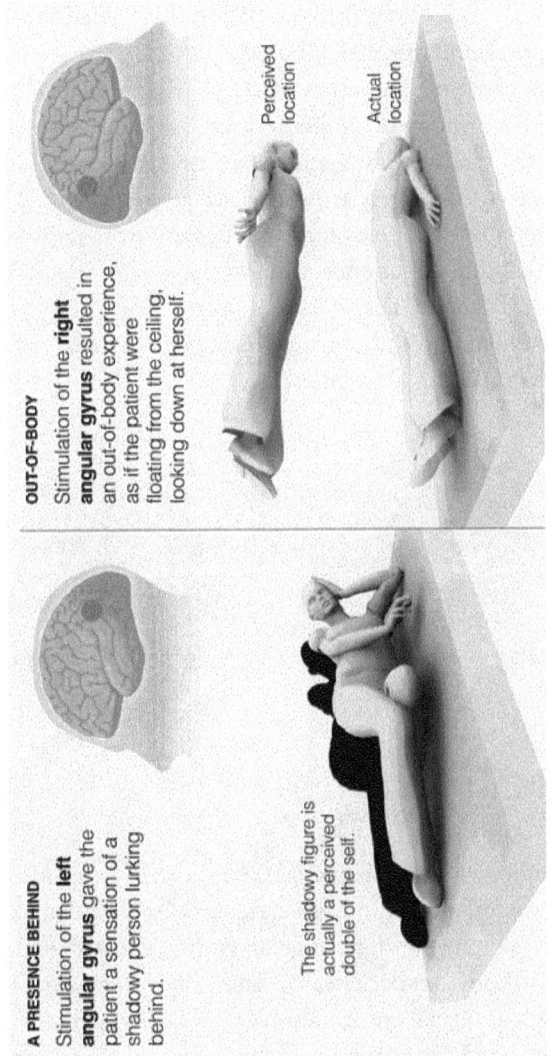

Figure 5.8 Dr. Blanke's study by stimulating the left and right angular gyrus

In most cases of NDEs, OBE is associated with the temporal and parietal lobe of the brain, especially the temporoparietal junction and the right angular gyrus. After a traumatic injury, these regions sometimes become malfunctional and start firing random gibberish neural signals. There is no specific region involved here. Rather there's a combination of brain areas that creates out-of-body experience when the brain is traumatized. We neuroscientists have been simulating those experiences in the controlled environment of the laboratory for a long time.

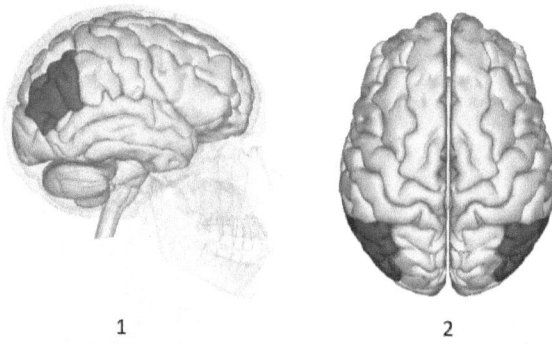

1   2

Figure 5.9 Angular gyrus (1. lateral view, 2. superior view)

Neurologist Olaf Blanke of Geneva University Hospital in Switzerland found that electrically stimulating one brain region, the right angular gyrus, repeatedly triggers out-of-body experiences. The right angular gyrus integrates visual information such as the sight of your body and information that creates the mind's representation of your body. This is based on balance and feedback from your limbs about their position in space.

With gentle stimulation, a woman, who could speak during an operation, felt she was falling or growing lighter. As the intensity increased she said *"I see myself lying in bed, from above."*

When, the left angular gyrus was stimulated, the patient reported a sensation of a shadowy figure lurking behind.

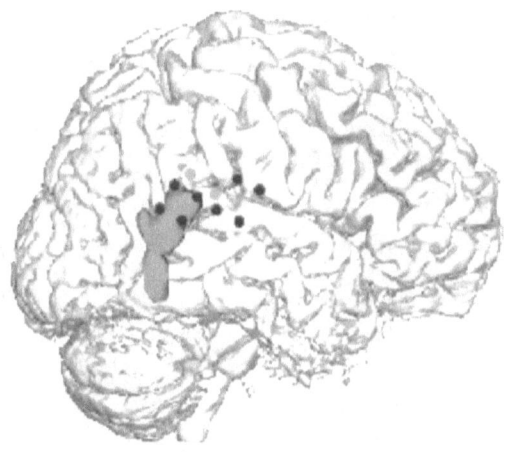

Figure 5.10 Lesions in the temporal lobe of an epileptic OBE experient

OBE doesn't necessarily involve near-death condition. It's all about the impaired brain regions that create OBE. And the impairment can be caused both by near-death episodes and by diseases like epilepsy. In many cases, OBE has been experienced by epileptic patients, as well as patients with circumstantial brain damage. In some epileptic OBE experients, lesions have been found in the temporal lobe. Some OBE patients have posttraumatic brain damage in the parietal lobe.

However, the right angular gyrus is the most involved brain region for out-of-body experience. Whereas, the temporal lobe is responsible for a wide array of religious/spiritual/mystical experiences (RSME).

Now comes the vivid higher dimensional perception and encounters of strange creatures, angels and diseased relatives. The rough reality is, these experiences where a person's soul or consciousness pops out of the head and visits a higher realm, are nothing but profound hallucinations of the human brain. Why does it seem so real that along with a layman even a smart neurosurgeon gets deluded? Because that's the point. The brain creates a virtual world of serenity so that the person can survive the traumatic episode.

Behind the creation of this inexplicable virtual reality, there is one specific gland at play. It's the very tiny pineal gland. And just like the old saying goes, good things come in small packages. The pineal gland is indeed a tiny wonder that allows a person in severe distress to wander around the spirit world. That's why it is sometimes considered as "the spirit gland".

The pineal gland is also falsely considered as the third eye, as you saw in the last chapter. Why false? Because, the characteristic of third eye is to sense the reality all around us, that cannot be sensed otherwise with our basic five sensory organs. Third eye is responsible of sensing the psi phenomena. And as you've seen in the

earlier chapters, psi info is sensed through the cerebro-
photonic field, which has almost nothing to do with the
pineal gland. So, the third eye is just the representation
of your mind's eye.

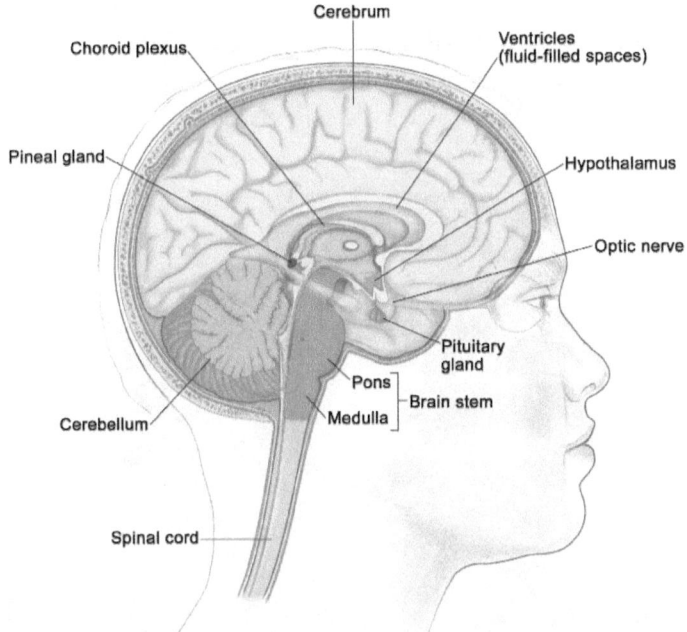

Figure 5.11 The tiny Pineal Gland (look how tiny it is!)

The pineal gland is unique in its solitary status within
the brain. All other brain sites are paired, meaning that
they have left and right counterparts; for example, there
are left and right frontal lobes and left and right
temporal lobes. As an unpaired organ deep within the
brain, the pineal gland remained an anatomical curiosity
for nearly two thousand years.

Interest in the pineal gland accelerated after it attracted Rene Descartes' attention. This seventeenth-century French philosopher and mathematician, who said, *"Je pense, donc je suis"* which means *"I think, therefore I am"*, wanted to explain the source of human thoughts. Introspection showed him that it was possible to think only one thought at a time. From where in the brain might these unpaired, solitary thoughts arise? Descartes proposed that the pineal gland, the only singleton organ of the brain, generated thoughts.

Figure 5.12 Rene Descartes

In addition, Descartes believed that the pineal's location, directly above one of the crucial byways for the cerebrospinal fluid, made its function of generating thoughts even more likely. The ventricles, hollow cavities deep within the brain, produce cerebrospinal fluid. This clear, salty, protein-rich fluid provides cushioning for the brain, protecting it from sudden jolts and bumps. It also carries nutrients to, and waste products away from, deep brain tissue.

In Descartes's time, the ebb and flow of the cerebrospinal fluid through the ventricles seemed perfectly suited for the corresponding movement of thoughts. If the pineal gland "secreted" thoughts into the cerebrospinal fluid, what better means for the "stream of consciousness" to make its way to the rest of the brain?

Descartes also had a deeply spiritual side. He believed that thinking, or the human imagination, was basically a spiritual phenomenon made possible by our divine nature, what we share with God. His idea was that our thoughts are expressions of, and proof for the existence of, our soul. Descartes believed that the pineal gland played an essential role in the expression of the soul:

"Although the soul is joined with the entire body, there is one part of the body [the pineal] in which it exercises its function more than elsewhere. . . . [The pineal] is so suspended between the passages containing the animal spirits [guiding reason and carrying sensation and movement] that it can be moved by them . . . ; and it carries this motion on to the soul. . . . Then conversely, the bodily machine is so constituted that whenever the gland is moved in one way or another by the soul, or for that matter by any other cause, it pushes the animal spirits which surround it to the pores of the brain."

Descartes thus proposed that the pineal gland somehow was the "seat of the soul," the intermediary between the spiritual and physical. The body and the spirit meet there, each affecting the other, and the repercussions extend in both directions.

As absurd as it may sound, it's actually because of Descartes' deeply spiritual claims, pineal gland attracted the attention of the scientific community. He was a philosopher and mathematician, with no fMRI or EEG in the seventeenth century. So, you cannot really expect an accurate scientific explanation from him. His ideas of the pineal gland were like baby steps towards a more vivid explanation.

We first read about the physical pineal gland in the writings of Herophilus, a third-century B.C. Greek physician from the time of Alexander the Great. Its name comes from the Latin "pineus", relating to the pine, pinus. This little organ is thus piniform, or shaped like a pinecone, no bigger than the nail of your pinkie finger.

The human pineal gland becomes visible in the developing fetus at seven weeks, or forty-nine days, after conception. The pineal gland of vertebrates develops from the neural crest cells of the roof of the diencephalon. In human beings it is deeply situated in the midline of the brain below the corpus callosum.

For many years physiologists considered the mammalian pineal gland the equivalent of the "brain's appendix". It was thought to be a residual, vestigial organ, a throwback to our early reptilian days, with no known role. This early idea changed when American dermatologist Aaron Lerner discovered melatonin in 1958. This and related findings called for the dawn of what might be called the era of the "melatonin hypothesis of pineal function".

Pineal gland is basically a light-sensitive organ that controls the wakefulness through melatonin. Melatonin regulates both seasonal and circadian rhythms. Darkness of the night induces secretion of melatonin which in turn controls the sleep pattern. It also regulates body temperature during sleep. So, in order to get a good night's sleep it is necessary that your environment is dark.

But, melatonin is not the compound of interest in this chapter. The most interesting and extraordinary compound that the pineal gland makes is DMT or N,N-Dimethyltryptamine, which by the way is a profound hallucinogenic compound. It's like a sibling compound of serotonin. The release of DMT in the pineal gland during any traumatic episode puts the brain into extraordinary mystical states.

Practically, when the entire body is under severe stress from a traumatic event, the pineal gland is the only mysterious organ that remains calm and actually simulates a virtual environment of peace and happiness to keep the person as much calm as possible. It has its own protective mechanism that gets highly alert during traumatic events. And obviously the most stressful period in a person's life is at the time of death.

Tremendous outpourings of stress-related hormones occur at this time, including the pineal-stimulating catecholamines adrenaline and noradrenaline which stimulate the pineal gland to produce more DMT, so that the last few moments of the person's life are not so stressful. And, since DMT is an intensely psychedelic compound, the last thing a dying person is conscious of, is being in an extraordinary realm of mysticism, which is created by the human brain with the profound influence of DMT.

DMT exists in all of our bodies and occurs throughout the plant and animal kingdoms. It is a part of the normal makeup of humans and other mammals; marine animals; grasses and peas; toads and frogs; mushrooms and molds; and barks, flowers, and roots.

Figure 5.13 DMT crystals

In humans DMT triggers such states of consciousness that can be interpreted as the common experiences of an afterlife. Many researchers investigated the influence of DMT over the human consciousness. Between 1990 and 1995 Dr. Rick Strassman, Clinical Associate Professor of Psychiatry at the University of New Mexico School of Medicine studied the effects of DMT by administering several hundred doses of DMT on volunteers.

The experiences that the volunteers had are similar to the vivid transcendental perceptions of near-death experients. One of the volunteers reported :

"It was very intelligent. It wasn't at all humanoid. It wasn't a bee but it seemed like one. It was showing me around the hive. It was extremely friendly, and I felt a warm sensual energy radiating throughout the hive. I decided it must be a wonderful thing to live in a loving and sensual environment such as that. It said to me that this was where our future lay."

There have been several other studies of the influence of DMT along the way. These studies revealed that there are several elements to the influence of DMT.

Visual, physical and auditory hallucinations are few of those basic elements. One volunteer stated :

"The room erupted in incredible neon colors, and dissolving into the most elaborate incredibly detailed fractal patterns that I have ever seen."

Another participant emphasized the dynamic complexity of the visual hallucinations :

"My visual field was consumed with disturbances, and they quickly escalated in intensity. There was all kinds of morphing,

bending, rippling, and breathing of objects when my eyes were opened. The entire room was crawling with beautiful geometric hallucinations."

Another feature of DMT is visiting other realities. This sometimes includes having contact with other sentient beings, which are often described by the volunteers as true or real experiences rather than hallucinations. Participants in various DMT studies described being transported to another world, reality, or universe, whereby one's ordinary waking experience was completely supplanted by what may be referred to as "DMT hyperspace". It is perhaps noteworthy that these realms were attributed greater veridicality than experiences associated with ordinary waking consciousness, which means, the reality perceived under the influence of DMT seems more real than the true everyday reality.

A specific feature of DMT realms interests us neuroscientists very deeply. Volunteers often reported that these realms were inhabited by entities that tended to fulfil the positive performative function of imparting knowledge about themselves and the universe. This theme is consistent with phenomenological reports of NDEs, which often include travelling to another place and meeting sentient beings, typically in the form of deceased persons. For example, in a study, one participant stated:

"I was greeted with a chrysanthemum in vivid limeish green and deep red, which then opened up into another plane, or dimension of

existence, or some type of parallel conscious living part of the universe."

Another participant reported interacting with entities that purportedly wished to impart knowledge:

"I realized there were some female entities swirling around me and one of them was actually showing me this mandala."

But perhaps the most extraordinary component of DMT-induced state of consciousness is Spirituality, learning or being taught about truths of the universe or one's self. Volunteers regularly gained insight into themselves and the nature of reality. Alternatively, this theme was simply expressed as a profound religious or spiritual experience. This result is consistent with the phenomenology of NDEs, which are often described as intensely religious, with the potential to exert a profound and enduring effect on the lives of the percipients. For instance, one volunteer recalled,

"Existing in this manner was the most joyful, religious, loving, happy, beautiful experience I have ever felt."

Another participant suggested that the DMT-induced experience was akin to enlightenment by referring to

"a higher intelligence that I now discovered I could tap into. It was like a religious experience, where I had a glimpse of what enlightenment could mean."

Furthermore, one participant purportedly gained metaphysical insights including knowledge pertaining to the structure of the universe:

"All I remember is I was shown how our physical three dimensional [sic] reality fits into the reality of the bigger universe that lies behind our three dimensional universe."

The DMT-induced hallucinogenic state is unlike any other psychedelic state. It feels so damn real, that even the real reality seems to fade away under the radiance of its veridicality. This is why, after going through a near-death experience many people have drastic transformation in their personality. But still, it is the craftsmanship of your brain, not of any spiritual entity.

This doesn't mean that I'm denying the existence of a supernatural spiritual entity, nor am I accepting it. As long as all the so-called paranormal near-death experiences can be explained in the juiciest biological way without incorporating any religious, spiritual or mystical component, it is totally unnecessary to waste time on worthless arguments out of sheer superciliousness. After all, if the extraordinary state of consciousness in a traumatic episode of a person's life gives him or her a few moments of inexplicable bliss and enables the person to survive the trauma, then I am not going to be an inhuman chauvinist to say that it is unscientific.

* * *

BOOK III

THE GOD PARASITE

REVELATION OF NEUROSCIENCE

INTRODUCTION

Here's what Science has proven through countless experiments:

God is a placebo.

God is a fundamental delusion.

God is a product of ignorance.

God is a product of fear.

God is a product of inability to explain.

God is a tactic of survival.

God is an experience just like love.

God is a fascination.

God is an illusory guide.

God is an evolutionary trick.

God is a product of weakness.

God is a mental state of the humans.

God is a neurological enigma.

God is often a symptom of an underlying disease.

Here's what all the religious scriptures express based on only the experiences of several primordial, ignorant, deluded and childish individuals:

God gave us our religions.

God demands us to kill people.

God will curse us if we don't convert.

God keeps record of our good and evil deeds.

God commanded me to change the world.

God told us to decapitate anyone who attempts apostasy.

God allows me to have more than one wives.

God commands us to stone inter-religious couples to death.

God can have as many as 16000 wives.

God teaches me to worship my husband.

God instructed me to prove my obedience to him by sacrificing my own son.

God will make my descendants as numerous as the stars in the sky and will give them all these lands.

God promised me that through my offspring all nations on earth will be blessed.

God commanded me to build an ark for my family and each pair of all animals, and let all other people on earth die when the planet is flooded.

God told me that I and my four brothers should marry
the same woman together.

Ultimately we can say that:

God is sociologically and psychologically constructed.

God is nothing but a neurological phenomenon.

God is the creation of the human brain.

I'm neither an atheist nor a theist. It's way too early to
get on either side. The existence of a Supreme Being as
a guiding force behind the universe, is irrelevant to this
book. Here I am going to reveal the molecular
interactions underneath the human experience and
perception of God and Divinity.

Being a neuroscientist, I don't get the luxury to get on
any side. If God is a delusion, which many atheist
authors suggest in their own way of unfounded rational
extremism, then there has to be a neurological basis for
that. Because all delusions are created in the brain. It is
only after investigating those neural underpinnings, that
we can confirm whether God is a delusion or something
much more inexplicable and elementary. The content of
this book is not meant to offend anyone. Rather this
book is a journey through the evolution of man while
he creates the concept of God based on his own needs
of daily survival.

Every supernatural phenomenon that humanity has
experienced so far since the primordial times, doesn't
have any unearthly divine intervention involved. The
entire belief system on a Supreme Creator was created

by primordial humans out of fear and ignorance, by means of illusions. Our primitive ancestors handed over their ideas of a supernatural sentient creator to us and as our brain size augmented over the period of evolution, we made the concept of God more sophisticated by means of our sophisticated hallucinations. And in the process we developed formal social structures of religion.

However, this book is not about finding the flaws in any religion or its scriptures, rather it is a voyage on which we shall discover that the entire industry of religion is born from fear, anxiety, ignorance, illusion, and neurological disorders. It is an evolutionary tactic for self-preservation, a coping mechanism.

Here, I don't make any attempt to take away anyone's religious beliefs whatsoever. This book is not a typical work of atheism. Rather it enables you to dive into the depth of the true biological foundation of the ever-astounding concept of God and Religions. Come with me to a fascinating land of questions, experiments and revelations, where we shall investigate, how and why the smartest species on earth constructed such a bizarre concept of a Supreme Omnipotent Entity.

* * *

CHAPTER 1
THE BIRTH OF GOD

"I belong to no religion. My religion is love. Every heart is my temple."

- Rumi

Mankind's first scientific achievement was to build tools and weapons out of rocks. Then they learnt to harness the power of fire. But somehow along the way a mysterious concept of supreme creator crept into the heart of primitive human. There were so many natural phenomena taking place all around the primitive man, which he couldn't explain by any means. It was way beyond his comprehension. Hence came fear. Fear of all natural manifestations of power got deep-rooted in his heart. So, man created his primitive religions out of fear by means of his illusions.

He worshiped every natural phenomenon that he could not comprehend. The observation of powerful natural forces, such as storms, floods, earthquakes, landslides, volcanoes, fire, heat, and cold, greatly impressed the expanding mind of man. These inexplicable phenomena of Mother Nature led him to come up with the most celebrated and apparently immortal term "Act of God".

At one time or another mortal man has worshiped everything on the face of the earth, including himself. He has also worshiped about everything imaginable in the sky and beneath the surface of the earth. The very first object that primitive man worshiped was stone. Different varieties of stones were sacred to the prehistoric humans. Stones first impressed early man as being out of the ordinary because of the manner in which they would so suddenly appear on the surface of a cultivated field or pasture. He failed to take into account either erosion or the results of the overturning of soil.

But the most profound influence was exerted by meteoric stones which primitive humans beheld coming down through the atmosphere in flaming grandeur directly from the heaven. The shooting star was inexplicably awesome to the early man, and he easily believed that such blazing streaks marked the passage of a spirit on its way to earth. No wonder humans were led to worship such phenomena, especially when they subsequently discovered the meteors. And this led to greater reverence for all other stones. This is not an act of God, this is just an act of childishly afraid evolving humans. However the disgusting reality is that the fetish for stones still remains in the heart of so-called civilized humans.

Following the cue of stone worship came something even worse, that is hill worship, and the first hills to be venerated were large stone formations. It became the custom to believe that the gods inhabited the mountains, so even high elevations of land were worshiped for this additional reason. As time passed, certain mountains were associated with certain gods and therefore became holy. This kind of delusion can easily be seen to this date in various ancient cultures throughout the world. For example, a specific hill in India, called "Govardhana" is considered to be a sacred form of the Supreme Lord Krishna (the Indian Casanova) in Hinduism and quite disgracefully is still worshiped every year the day after Diwali (the festival of lights).

In the ancient Indian text "Vishnu Purana" legends have it that the ancient people of Gokul used to worship Indra, the God of thunder and rain (like our beloved Thor). Krishna told them that they should stop worshiping Indra and worship more natural forms of the Supreme Creator, such as the hill Govardhana. The villagers were so convinced that they immediately stopped their arrangements for worshiping Indra and prepared to worship the hill Govardhana instead.

Figure 1.1 Lord Krishna lifting the Govardhana hill

Indra was damn furious and invoked clouds in the sky and schemed to flood the region with rains lasting for seven days and seven nights. Lord Krishna came forward to ensure the villagers' security and after performing worship and offering prayers to Mount Govardhana, he lifted it as an umbrella, on the little finger of his right hand, so that everyone could take shelter under it. After this event, Lord Krishna was also known as Giridhari or Govardhandhari (one who lifts Mount Govardhana). Since then the hill Govardhana is still being worshiped by the so-called civilized human beings of that region.

Even a toddler can tell that there was just heavy rainfall at some point of time and apparently the primitive villagers took shelter in the caves of the hill. And as the hill saved them from the disastrous rainfall they carried on worshiping it.

They were primitive ignorant villagers, so it was natural for them to worship anything that made their survival possible. But the most disgraceful thing is that people are still carrying out those rituals of the primitive period by identifying them with the later evolved concepts of good spirits and deities. If you walk around the streets of India, you'll find out that if you just pick up a stone the size of a volleyball and place it under a big tree by the roadside with some red 'kumkuma' (a powder usually red in color used for social and religious markings mainly in India) on it, in a little while people would start to visit the stone with utmost faith in an intention of worshipping it and give offerings to it. Good way to run a business, don't you think!!!

Figure 1.2 The Grand Mosque in Mecca

Figure 1.3 Left: Black Stone on the eastern corner of Kaaba, Right: Black stone as Shiva Lingam

Likewise the Black Stone became a highly venerated symbol of the Islamic Tradition. It is the eastern cornerstone of the Kaaba, the ancient stone building, located at the center of the Grand Mosque in Mecca, Saudi Arabia. It is revered by the Muslims as an Islamic relic which, according to Muslim tradition, dates back to the time of Adam and Eve. And actually the same kind of stone is worshiped as a representation of Lord Shiva in Hinduism. It is called Shiva Lingam.

There was no end to primitive man's childishness. Prehistoric humans even feared plants and trees. Therefore came the cults of plant and tree worship. They worshiped them because of the intoxicating liquors which were derived therefrom. Primitive man believed that intoxication rendered one divine. There was supposed to be something unusual and sacred about such an experience. Even in modern times alcohol is known as "spirits". For example, in Shamanic cultures, Datura stramonium, a plant in the Solanaceae (nightshade) family is still considered to be sacred because of its highly hallucinogenic properties.

Tree worship still takes place among the oldest religious groups. Trees were perceived as auspicious symbols, for which wedding ceremonies of the early clans and tribes used to take place under the trees. Many plants and trees were venerated because of their real or fancied medicinal powers. The savage believed that all chemical effects were due to the direct activity of supernatural forces.

Different tribes and clans had different supernatural perceptions about the spirit of trees and plants. Some trees were dwelt by friendly spirits while others were home to quite cunning and cruel spirits. The Finns believed that most trees were occupied by kind spirits. The Swiss long mistrusted the trees, believing they contained tricky spirits. The inhabitants of India and eastern Russia regard the tree spirits as being cruel.

Ficus benghalensis, the Indian Banyan tree till this date is conceived to be home for cruel spirits in India. The Patagonians still worship trees, as did the early Semites. Long after the Hebrews ceased tree worship, they continued to venerate their various deities in the groves. In China, there once existed a universal cult of the tree of life. But the most profound fetish for trees is found in the Bible. In the biblical reference from the Book of Genesis chapters 2 and 3, we find the description of the Garden of Eden where there were two extraordinary trees named the "tree of knowledge" and the "tree of life" among all others.

Bible : Genesis 2

By the seventh day God had finished the work he had been doing; so on the seventh day he rested from all his work. Then God blessed the seventh day and made it holy, because on it he rested from all the work of creating that he had done. This is the account of the heavens and the earth when they were created, when the LORD God made the earth and the heavens.

Now no shrub had yet appeared on the earth and no plant had yet sprung up, for the LORD God had not sent rain on the earth and there was no one to work the ground, but streams came up from the earth and watered the whole surface of the ground.

Then the LORD God formed man of dust from the ground, and breathed into his nostrils the breath of life; and man became a living being.

Now the LORD God had planted a garden in the east, in Eden; and there he put the man he had formed. The Lord God made all kinds of trees grow out of the ground - trees that were pleasing to the eye and good for food. In the middle of the garden were the tree of life and the tree of the knowledge of good and evil.

......

The LORD God took the man and put him in the Garden of Eden to work it and take care of it. And the LORD God commanded the man, "You are free to eat from any tree in the garden; but you must not eat from the tree of the knowledge of good and evil, for when you eat from it you will certainly die.

The LORD God said, "It is not good for the man to be alone. I will make a helper suitable for him." Now the LORD God had formed out of the ground all the wild animals and all the birds in the sky. He brought them to the man to see what he would name them; and whatever the man called each living creature, that was its name. So the man gave names to all the livestock, the birds in the sky and all the wild animals. But for Adam no suitable helper was found.

So the LORD God caused the man to fall into a deep sleep; and while he was sleeping, he took one of the man's ribs and then closed up the place with flesh. Then the LORD God made a woman from the rib he had taken out of the man, and he brought her to the man.

The man said, "This is now bone of my bones and flesh of my flesh; she shall be called 'woman,' for she was taken out of man." That is why a man leaves his father and mother and is united to his wife, and they become one flesh. Adam and his wife were both naked, and they felt no shame.

Bible : Genesis 3

Now the serpent was more crafty than any of the wild animals the LORD God had made. He said to the woman, "Did God really say, 'You must not eat from any tree in the garden'?"

The woman said to the serpent, "We may eat fruit from the trees in the garden, but God did say, 'You must not eat fruit from the tree that is in the middle of the garden, and you must not touch it, or you will die.'" "You will not certainly die," the serpent said to the woman. "For God knows that when you eat from it your eyes will be opened, and you will be like God, knowing good and evil."

When the woman saw that the fruit of the tree was good for food and pleasing to the eye, and also desirable for gaining wisdom, she took some and ate it. She also gave some to her husband, who was with her, and he ate it. Then the eyes of both of them were opened, and they realized they were naked; so they sewed fig leaves together and made coverings for themselves.

Figure 1.4 Garden of Eden

Then the man and his wife heard the sound of the LORD God as he was walking in the garden in the cool of the day, and they hid from the LORD God among the trees of the garden. But the LORD God called to the man, "Where are you?" He answered, "I heard you in the garden, and I was afraid because I was naked; so I hid."

And he said, "Who told you that you were naked? Have you eaten from the tree that I commanded you not to eat from?" The man said, "The woman you put here with me--she gave me some fruit from the tree, and I ate it." Then the LORD God said to the woman, "What is this you have done?" The woman said, "The serpent deceived me, and I ate."

So the LORD God said to the serpent, "Because you have done this, "Cursed are you above all livestock and all wild animals! You will crawl on your belly and you will eat dust all the days of your life. And I will put enmity between you and the woman, and between your offspring and hers; he will crush your head, and you will strike his heel." To the woman he said, "I will make your pains in childbearing very severe; with painful labor you will give birth to children. Your desire will be for your husband, and he will rule over you.

To Adam he said, "Because you listened to your wife and ate fruit from the tree about which I commanded you, 'You must not eat from it,' "Cursed is the ground because of you; through painful toil you will eat food from it all the days of your life. It will produce thorns and thistles for you, and you will eat the plants of the field.

Adam named his wife Eve, because she would become the mother of all the living. The LORD God made garments of skin for Adam and his wife and clothed them. And the LORD God said, "The man has now become like one of us, knowing good and *evil. He must not be allowed to reach out his hand and take also from the tree of life and eat, and live forever."*

So the LORD God banished him from the Garden of Eden to work the ground from which he had been taken. After he drove the man out, he placed on the east side of the Garden of Eden cherubim and a flaming sword flashing back and forth to guard the way to the tree of life.

It was the early juvenile mind of man that encouraged him to come up with this kind of illusory perceptions of the natural phenomena. It definitely didn't take only 7 days to create the universe and all the creatures in it, rather it took millions of years of evolution for our planet to be in the way it is now with all its creatures. So, creationism is total poppycock and it doesn't go along with the reality. As humans evolved, they discovered new techniques of worship while their roots of fearful illusory faith remained way back at the primordial times.

Primordial man had to live on this planet along with all other animals. Many of those animals were actually superior to humans in strength and ferociousness at that prehistoric time. This brought fear, which brought worship of animals. There were some animals which were absolutely friendly and provided means for survival, like milk from the cow. Hence came gratitude that again brought blind superstition of worshiping cows, which is still pretty robust among the Hindus. But it doesn't end here, we have a long list of superstitious garbage remaining to be analyzed. For instance, the

Hindus in the remote villages of India still maintain friendly relations with their house snakes.

The Hebrews worshiped serpents down to the days of King Hezekiah. The Chinese worship of the dragon is a survival of the snake cults. The wisdom of the serpent was a symbol of Greek medicine and is still employed as an emblem by modern physicians. The art of snake charming has been handed down from the days of the female shamans of the snake love cult, who, as the result of daily snake bites, became immune. They even became genuine venom addicts and could not get along without this poison.

Quite ridiculously, the ancients even believed that all winds were produced by the wings of birds and therefore they both feared and worshiped all winged creatures.

But perhaps the most celebrated worship by the ancients, that still goes on with all the glory is the worship of elements: earth, air, water and fire. The primitive people venerated rivers. Many believed that by bathing in a sacred river, they would get rid of all their sins. Even now in Mongolia flourishes an influential river cult. Baptism became a religious ceremonial in Babylon, and the Greeks practiced the annual ritual bath. It was easy for the ancients to imagine that the spirits dwelt in the bubbling springs, gushing fountains, flowing rivers, and raging torrents.

Moving waters vividly impressed these simple minds with beliefs of spirit animation and supernatural power. Sometimes a drowning man would be refused rescue for fear of offending some river god. In India, the river Ganges is still considered to be holy by the Hindus. It is worshiped as goddess Ganga in Hinduism. Millions of people gather every year in the holy city of Varanasi, in order to take a dib in the sacred Ganges to get rid of their earthly sins. But don't even dare to take a bath in that polluted river yourself, because it's filled with dangerous parasites and even people's poop. Yet, it is the naive Indians' blind superstition that allows them to live in the prehistoric age while perceiving such river water as elixir of life.

The ancient Bedouins believed that a nature spirit produced the sand whirls, and even in the times of Moses belief in nature spirits was strong enough to insure their perpetuation in Hebrew theology as angels of fire, water, and air.

Clouds, rain, and hail have all been feared and worshiped by numerous primitive tribes at some period of time or other. Windstorms with thunder and lightning had deep impact over the early man. He was so impressed with these elemental disturbances that thunder was regarded as the voice of an angry god. The worship of fire and the fear of lightning were linked together and were widespread among many early cultures.

Worship of elements concludes the veneration of earthly natural objects. Now comes the unearthly ones, which means the heavenly bodies like the sun, moon and stars. In many cultures the stars were regarded as the glorified souls of great men who had departed from the life in the flesh. Worship of heavenly bodies began with the worship of moon and then came worship of the sun.

With all these manners of veneration primordial humans wanted to deal with their fear for the manifestations of natural powers. This ingrained fear and ignorance about the natural forces, gave rise to the incredible mental feature of spirituality. This feature eventually ended up being a blessing for the species, as it started working as a natural anti-depressant for the human mind. And upon the seeds of spirituality, relatively modern humans developed formal structures of religion to sustain that spirituality.

Before the birth religions, the more primeval knack of the ignorant and scared human minds was to perceive a supernatural air all around them. And this mystical knack of the mind led scientists to develop various terms – animism, fetishism and totemism. Even though they are not directly connected to modern religions, they prepared the soil of the human mind to give birth to spirituality, which later gave rise to various mystical experiences of utmost religious significance.

During the nineteenth century terms like fetishism, totemism and animism were developed by European social scientists for describing what we can call the original religions of humanity. They had their own rituals, beliefs and regulations, which were ought to be followed by generations after generations, much like our modern religions. However, at the base of all these original religions, there were the common elements of ignorance, fear and anxiety, while the modern religions are the creation of much more complicated neural processes.

The original religions were the products of the savages' inability of assessing the meaning and value of material objects. They were the most primitive forms of spirituality. In terms of this early spirituality there was really not much difference between humans and dogs. We can understand the primitive spirituality by observing the psychology of dogs.

In his popular survey of human evolution, The Origin of Civilization and the Primitive Condition of Man, John Lubbock explained that religion originated as the result of the primitive tendency to attribute animation to inanimate objects, which is known as animism. However, you'll read in the coming chapters that religion has much more than just animism at its basis. Anyways, Lubbock observed, *"dogs appear to do the same"*. As Lubbock's friend and mentor, Charles Darwin thought that religion could be explained in terms of dog behavior.

Like Lubbock, Darwin observed that dogs characteristically attributed life to inanimate objects. His dog's attention to a parasol blowing in the wind, suggested to Darwin that the animal assumed that objects were alive. In this animal psychology, therefore, nineteenth-century theorists had a basis for understanding animism as the primitive or savage propensity to attribute animation to inanimate objects.

So, technically when mystics proudly proclaim that spirituality is what makes us superior to the animals, they are actually boasting their false vanity. Because, if spirituality makes a species special, then even the dogs would be called as spiritual. So, basically we humans are not much different from dogs on this matter. There is nothing supernatural about spirituality after all.

Our primitive ancestors looked upon the whole nature as alive, as possessed by something. They attributed life and spirit to even the inanimate objects. This is what anthropologist Sir Edward B Tylor termed as "Animism". The savages' primitive beliefs in the Spiritual Beings led to the modern sophisticated sense of spirituality that most of the modern humans possess.

The notion of spirituality is wider, less concrete and less institutionally bound than religiosity. Spirituality is the heart of all religions, yet it can prevail beyond the bounds of religious structures, that's why many non-religious people consider themselves to be spiritual.

Now we need to find out, how exactly the term "spirit" emerged?

The primitive ancestors of humanity were unable to distinguish between dreams and waking consciousness. To their childish mind, dreams were just as real as any other waking experience. When they dreamed about deceased friends or relatives, they assumed that the dead were still alive in some spiritual form. It was all very real to the savage who would awaken from such dreams reeking with sweat, trembling, and screaming. Out of dreams, therefore, evolved the doctrine of souls and

196

other spiritual beings, in general, a doctrine that was rational to the savages, even though it was a childish philosophy enveloped in intense and inveterate ignorance.

This doctrine grew stronger aided by the biological instinct of self-preservation as it effectively helped to antidote the fear, anxiety and uncertainty of death. In addition, the involuntary physical phenomenon of sneezing fueled the idea of "soul" or "spirit". Sneezing was not originally an arbitrary and meaningless custom, but the working out of a principle.

Anthropologist E. B. Tylor derived the ethnographic facts that the Zulus thought their deceased ancestors caused sneezing. Sneezing reminded the Zulus to name and praise their ancestors. They believed that the ancestors entered the bodies of their descendants when they sneezed. Ritual specialists, such as Zulu diviners, regularly sneezed as a ritual technique for invoking the spiritual power of the ancestors. Tylor concluded that these Zulu concepts and practices, were remnants of a prehistoric era in which sneezing was not merely a physiological phenomenon, but was still in the theological stage. Many of the savages also looked upon sneezing as an abortive attempt of the soul to escape from the body. Eventually, sneezing got accompanied by some common social expression, such as "bless you".

As a foundational element of spirituality, animism encompasses the belief that there is no separation between the spiritual and physical or material world.

This belief expresses that souls or spirits exist, not only in humans, but also in inanimate objects such as plants, rocks, mountains, rivers and other entities of the natural environment. Along came the terms "fetishism" and "totemism".

The term "fetishism" was derived from the Portuguese word "feitiço", referring to nefarious instruments of magic and witchcraft. A fetish is an object believed to have supernatural powers. Initially, the Portuguese developed the concept of fetishism, referring to the objects used in religious cults by West African natives.

Contemporary Portuguese feitiço translates as charm, enchantment, juju or abracadabra, or more potentially offensive witchcraft, witchery, conjuration or bewitchment. But today, fetishism basically means "obsessive fascination". Fetishism is also characterized as a disorder when there is a pathological assignment of sexual fixation, fantasies or behaviors toward an inanimate object, frequently an item of clothing such as underclothing or a high-heeled shoe or to nongenital body parts such as the foot, armpit etc.

In such sexual fetishes, only through the use of that specific object can the individual obtain sexual gratification. The fetishist usually holds, rubs or smells the fetish object for sexual gratification or asks their partner to wear the object during sexual encounters. Fetishism is a more common occurrence in males, and the causes are not clearly known yet. Fetishism falls under the general category of paraphilias, i.e. abnormal or unnatural sexual attractions.

Sigmund Freud conjectured on the symbolic origins of sexual fetishes but did not explore the meaning of fetishes systematically as he did dream contents. He treated fetishism as a deviation from the normal sexual aim of copulation leading to the release of sexual tension. In 1927, Freud published his paper on "Fetishism", in which he illustrated the analytic meaning and background of sexual fetishism.

Most of Freud's ideas of psychoanalysis emerged from his highly sexual perspective. In his paper on "Fetishism" he writes:

In every instance, the meaning and the purpose of the fetish turned out, in analysis, to be the same. It revealed itself so naturally and seemed to me so compelling that I am prepared to expect the same solution in all cases of fetishism. When now I announce that the fetish is a substitute for the penis, I shall certainly create disappointment; so I hasten to add that it is not a substitute for any chance penis, but for a particular and quite special penis that had been extremely important in early childhood but had later been lost. That is to say, it should normally have been given up, but the fetish is precisely designed to preserve it from extinction. To put it more plainly: the fetish is a substitute for the woman's (the mother's) penis that the little boy once believed in and - for reasons familiar to us - does not want to give up.

But, these are much modern forms of fetishism, evolved in the modern times of civilization. If we just turn the clock back to the primitive times, we'll notice that fetishism really had one heck of a hold over the primitive people. And this magical thinking along with Animism and Totemism influenced the evolution of modern Spirituality. The world of mystical philosophy implies that Spirituality makes us different from the animals. But in fact, it's quite the opposite. We may have nurtured our perception of spirituality over time and have made it modern, but deep down, our primordial instincts of spirituality are just as same as the dogs'.

Now let's analyze another foundational element of spirituality, i.e. Totemism. The term "totemism," according to the Scottish ethnographer John Ferguson McLennan, referred to communal alliances under the sign of an animal or an object that combined fetishism with exogamy, mixing the inability to evaluate materiality with regulations governing sexuality. In simple words, Totemism is a belief that each human has a spiritual or mystical kinship with a spirit-being, called as a "totem", such as an animal or plant. The totem is thought to interact with a given kin group or an individual and to serve as their emblem or symbol. Therefore the totem becomes a highly venerated figure of that specific kin group. This primitive idea is still alive in many cultures around the world in some form or another. For example the many traditional Hindus still have their totemic belief in their "Kuldevata", that means *"deity of the kin group"*.

Totemism was really a fantastically bizarre idea. And technically it has given rise to many of the modern governing regulations of sexuality. The term "totem" is derived from the Ojibwa word "ototeman", meaning "one's brother-sister kin". The grammatical root, 'ote', signifies a blood relationship between brothers and sisters who have the same mother and who may not marry each other.

The English word totem was introduced in 1791 by a British merchant and translator who gave it a fabricated meaning. He believed that the term "totem" designated the guardian spirit of an individual or a kin group, that appeared in the form of an animal. The Ojibwa clans did indeed portray such idea by wearing animal skins. The Ojibwa named their clans after those animals that lived in the area in which they lived and appeared to be either friendly or fearful.

The first accurate report about totemism in North America was written by a Methodist missionary, Peter Jones, himself an Ojibwa, who died in 1856 and whose report was published posthumously. According to Jones, the Great Spirit had given toodaims ('totems') to the Ojibwa clans, and because of this act, it should never be forgotten that members of the group are related to one another and on this account may not marry among themselves. Thus evolved Exogamy. Hence in modern times sexual intimacy within family is perceived as a taboo. From this perspective the evolving mind of the humans created the concept of "Incest".

However, there is an evolutionary benefit of the concept of Incest. Marriage outside the family and even outside the totem groups brought diversity in the genetic blueprint of the progeny. And diversity has always been a very important element in the evolutionary process of all species on earth for the development of new genetic traits. Hence exogamy proved beneficial for our species.

In this 21st century we are living in our classy apartments in the mist of smart technological wonders. So our today's world is really what may be called a world of an advanced civilization. Around us, we have our technological extensions. Cars are the extension of human legs; our smartphones are the extension of our ears, mouth and fingers. Flights are the human extension of wings. So, technically we have lost contact with nature, or to be more specific, with wild nature. Modern Homo sapiens don't live in the jungle any more. So we no more have to compete with our fellow animal species.

Now look at the conditions of your primitive ancestors. They were savages who had only raw natural resources at their disposal in order to survive each day. Mother Nature was their nest. While living in the unsophisticated natural environment their natural tendency was to postulate intimate connections between human life and natural beings or objects around them.

However, as time passed the primitive idea of the spirit world evolved as well, with a denser perception of something beyond the human world. In time, the modern concept of God had its true inauguration by the complex neural network of the three pounds spongy organic material known as the human brain.

* * *

CHAPTER 2
UPBRINGING OF GOD

Evolution is the most common element of life on planet earth. It is in fact the bed-rock of biology. After being apes for a long time, we have evolved into so-called civilized humans. But so did our God. Throughout the period of evolution somehow this specific trait of the human species not just survived, it has evolved along with the human brain. As time passed, the evolving mind of the primitive humans started to create God within themselves. They began to formulate the concept of God with the characteristics they had. They created God in their own image. Hence occurred the arrival of modern religions.

When religion once evolved beyond the primitive nature of worship, it acquired roots of spiritual and more of a human origin. This way the various Gods we have today are just the spiritual representation of the humans. Humans wanted something more humanlike to worship, so they created various superhuman personalities. Quite naturally all the basic characteristics of the primordial humans became the characteristics of the Gods that we have today. Some of those characteristics are really beneficial to the mankind philosophically, while some are downright harmful. Man created the image of God by means of fear and various illusory spiritual encounters which never actually happened other than within the brain. Throughout history prophets have been worshiped as the messengers of God, because apparently they talked to God. The illusions of those prophets' became the illusion of the entire humanity. These illusions were not just the childish illusions of primordial man, rather, they vividly took place within the complex neural network of the brain. This means that the brain had the neural capacity to work on a higher level than the primitive ancestors. The intense urge of man to perceive God in his own image, compelled the evolving brain to create encounters of a spiritual kind.

Yes you heard right….. the Gods we have today were actually created by the human brain. It is irrelevant to the book whether there is any real Supreme Sentient Being out there somewhere. But the Gods that mankind has so far experienced were unquestionably, extreme hallucinations of the brain. How the human brain creates God and all other spiritual experiences will be illustrated in a later chapter.

So, the fundamental mantra of the fundamentalists that *"God has created the humans"* is total poppycock. It's rather quite the opposite: *"Man created God"* due to his own selfish desires.

Early man was naïve and whoever came up with the story that he or she had encountered a supreme personality, was comprehended as a prophet or messenger of God. This way each prophet formed a separate religion with a group of early naïve humans craving for answers to the complexity of life. The prophets themselves were deluded and they made all their followers deluded as well.

Those evolving people regarded all unusual indivudals as superhumans, and even sometimes literally worshiped them with utmost respect. Lunatics, epileptics and other mentally ill individuals were often venerated by the naïve humans as they believed that such abnormal behaviors were signs of God dwelling in them. And quite unfortunately even today such abnormal behaviors are considered to be the signs of "possession" by a demon or deity spirit. I guess the churches needed some strong basis to carry on their flourishing business of exorcism. The situation is worst in India. I can't tell you how many mentally ill people have been killed by the masses with the accusation of being possessed by witches. Again, at the same time, there are specific religious sites in India where people still visit in an intention to behold the possession of a deity over some individual. It's a disgrace to say that even so-called educated people visit those places with the belief that they would encounter their deity spirit. Even though it is scientifically revealed today that demonic possession is nothing but epileptic seizures, people still sometimes like to believe their primordial instincts.

If you are an atheist, then don't even for a second think that it is the fault of those believers. It is absolutely not. To speak simply, it is how our brain has evolved. God is hardwired within our inner neural circuitry. I'll show you how, in the next chapter.

Fear gives birth to a savior. If you are fearless, you don't need an illusory savior. This is exactly what happened to our ancestors. They were ignorant, naïve, afraid and unable to comprehend the complexities of life. Man's earliest prereligious fear of the natural forces gradually became religious in nature and got personalized and spiritualized. Eventually he learnt to say *"God works in a mysterious way"*.

Those who created Gods in the way we know now, were savages. They lived in the jungle and roamed around the deserts. Anxiety and fear were the natural states of the savage mind. The struggle for life was excruciatingly painful. But again pain and suffering were and still are essential for the evolution of a species. But any species that is not strong enough to survive the dread of living gets extinct from the face of earth. Faith in the existence of a supernatural being was man's means to deal with the miseries and fear.

Mankind has always been a slow learner. The primitive people were limited in their ability of logicality and intellectual associations. They were savages with an uneducated and totally unsophisticated mind. Naturally they were inclined to believe in whatever they deemed best for themselves. They believed in whatever was in their immediate or remote interest. Therefore self-interest largely did and still does obscure logic. Attributing things difficult of comprehension to supernatural causes is nothing less than a lazy and convenient way of avoiding all forms of intellectual hard work.

Another inexplicable shock to evolving man was death. Death was filled with perplexing mysteries to the savages. In the human mind there was already the nebulous concept of a hazy and unorganized spirit world. It was a domain full of inexplicable stuff, and death was added to this long list of unexplained phenomena.

All human diseases and natural deaths were at first believed to be due to spirit influence. As humans evolved and more complex systems of so-called theology erupted, they still ascribe death to the action of the spirit world.

Dreams of a dead member of the tribe gave rise to the concept of soul. The simultaneous dreaming about a departed chief by several members of his tribe seemed to constitute convincing evidence that the old chief had really returned in some form. It was all very real to the savage who would wake up from such dreams rancid with sweat, trembling and screaming. This lead to the rise of the concept of afterlife. It began effectively to remedy the fear of death associated with the biological instinct of self-preservation.

Early man was also much concerned about his breath. The breath of life was regarded as the one phenomenon which differentiated the living and the dead. He believed the breath could leave the body, and his vivid dreams consisting all sorts of queer dream contents convinced him that there was something immaterial about a human being. This was the origin of the most primitive idea of the human soul. Eventually the savages were convinced that the breath without the body tantamount to a spirit or a ghost.

Primitive humans could not comprehend the vastness of infinity and eternity, so as a trick of self-preservation they came up with the perception of survival of the soul after death and its recurring incarnations.

Eventually this belief of the ancient ancestors in transmigration and reincarnation greatly inspired the ancient sages to form the contents of scriptures that later became the foundation of modern religions.

The ancient Indian sage Vyasa, a central and highly venerated figure in most Hindu traditions penned the concept of immortality of the soul in Bhagavad Gita, a part of the Hindu epic Mahabharata written in the 3rd millennium BC.

Bhagavad Gita: Chapter 2

Verse 22

> vasamsi jirnani yatha vihaya
> navani grhnati naro parani
> tatha sarirani vihaya jirnany
> anyani samyati navani dehi

Translation:

As a person puts on new garments, giving up old ones, the soul similarly accepts new material bodies, giving up the old and useless ones.

Verse 23

> nainam chidanti sastrani
> nainam dahati pavakah
> na caiman kledayanty apo
> na sosayati marutah

Translation:

The soul can never be cut to pieces by any weapon, nor burned by fire, nor moistened by water, nor withered by the wind.

So eventually the immortality of the soul became a foundational belief in modern religions. In many religious traditions worldwide it is believed that the individual soul comes from the supreme soul. Later on, the idea of heaven and hell occurred regarding the destiny of good soul and sinful soul.

A little fun fact: the savage looked upon sneezing as an abortive attempt of the soul to escape from the body. Being awake and on guard, the body was able to thwart the soul's attempted escape. Later on, sneezing was always accompanied by some religious expressions, such as *"God bless you"* or just *"bless you"*.

In due course of time, the breath got deeply associated with soul. The early tribal people believed that the qualities of the soul could be imparted or transferred by the breath. The brave tribal chief would breathe upon the newborn child, thereby imparting courage. Among early Christians the ceremony of bestowing the Holy Spirit was accompanied by breathing on the candidates. It was long the custom of the eldest son to try to catch the last breath of his dying father.

All these ideas of the primitive religion prepared the soil of the human mind, by the rough and merciless forces of fear and ignorance to sow the seeds of modern Gods and their background religions. And the human brain slowly created the sophisticated forms of modern Gods out of man's innate urge to have a guardian figure.

* * *

CHAPTER 3

ENCOUNTERS WITH GOD

"You dance inside my chest, where no one sees you, but sometimes I do, and that sight becomes this art."

\- Rumi

From the fertile soil of the primitive human mind grew the trees of modern religions. The very rudimentary nutrients for those trees at their early days came from the human brain itself. When the primitive days of mankind were over, some people out of the masses all over the globe started to have vivid hallucinations of a supreme sentient being. These so-called encounters with God powered the arrival of modern scripture based religions.

So basically all the philosophical teachings in the religious scriptures some of which are good and some are total garbage came from the human mind, not from any God in the sky who might be rolling his eyes over us.

Abraham, Moses, Muhammad, Jesus, Vyasa, Buddha, Joseph Smith, Nanak, Joan of Arc and many others were just ordinary human beings who were craving for philosophical guidance so damn much that their own brain created the virtual environment of meeting God and receiving messages from him. And the messages that they received were just the instinctual responses of their own mind. That's why no scripture is totally accurate or beneficial. All of them are filled with both goodness and flaws. Because they were the creation of humans.

So let's talk about some of those encounters that eventually gave rise to modern religions. Abraham for instance claimed to talk to God. The entire foundation of Judaism and Hebrew Bible was built upon Noah, Abraham and Moses's encounters with God. Likewise the oldest religion on earth that is Hinduism was built upon the foundation of the ancients Indus Valley texts, Vedas, Upanishads Puranas and many others. Hinduism is the accumulated treasury of spiritual laws experienced by different ancient sages in different times. Here the ancient Indian sages claim to have received the spiritual laws from a supreme sentient being. Same goes for Mohammad for founding Islam, Buddha for Buddhism, Nanak for Sikhism and many others.

Apparently Noah got a phone call from God commanding him to build an Ark and fill it with all sorts of living creatures, because the earth was about to be flooded. The same kind of event is mentioned in the ancient Indian chronicle Matsya Purana. In this story Lord Vishnu commanded King Manu to build a gigantic boat and fill it with all kinds of animals and plant specimens to escape the Great Deluge.

Figure 3.1 Noah's Ark on Mount Ararat

According to the Matsya Purana, the Matsya (fish) Avatar of Vishnu first appeared as a small carp, to Manu, while he was washing his hands in a river flowing down the Malaya Mountains in his land of Dravida.

Figure 3.2 Matsya Avatar of Lord Vishnu during The Great Deluge

The little fish asked the king to save Him, and out of compassion, he put it in a water jar. It kept growing bigger and bigger, until the king first put it in a bigger pitcher, and then deposited it in a well. When the well also proved insufficient for the ever-growing fish, the King placed it in a reservoir that was 16 miles in height above the surface and on land, as much in length, and 8 miles in breadth. As it grew further, the king had to put the fish in a river, and when even the river proved insufficient, he placed it in the ocean, after which it nearly filled the vast expanse of the great ocean.

It was then that Vishnu, revealing himself, informed the king of an all-destructive deluge which would be coming very soon. The king built a huge boat which housed his family, the seven sages, nine types of seeds and animals to repopulate the earth, after the deluge would end and the oceans and seas would recede. At the time of deluge, Vishnu appeared as a horned fish and Shesha Naga (king of all serpent deities) appeared as a rope, with which the king fastened the boat to the horn of the fish.

The boat was perched after the deluge on the top of the Malaya Mountains. After the deluge, Manu's family and the seven sages repopulated the earth, just like in Hebrew Bible Noah's family repopulated the earth.

About four hundred years after Noah's death, his descendant Abraham was called by God. The entire narrative of Abraham revolves around abundant progeny and authority of lands. Abraham experienced the two most crucial promises of the religious history of mankind that changed the world forever.

Bible : Genesis 12 : 1-7

The LORD had said to Abram, "Go from your country, your people and your father's household to the land I will show you.

"I will make you into a great nation, and I will bless you; I will make your name great, and you will be a blessing.

I will bless those who bless you, and whoever curses you I will curse; and all peoples on earth will be blessed through you."

So Abram went, as the LORD had told him; and Lot went with him. Abram was seventy-five years old when he set out from Harran.

He took his wife Sarai, his nephew Lot, all the possessions they had accumulated and the people they had acquired in Harran, and they set out for the land of Canaan, and they arrived there.

Abram traveled through the land as far as the site of the great tree of Moreh at Shechem. At that time the Canaanites were in the land.

The LORD appeared to Abram and said, "To your offspring I will give this land." So he built an altar there to the LORD, who had appeared to him.

Bible : Genesis 22 : 1-18

Some time later God tested Abraham. He said to him, "Abraham!" "Here I am," he replied.

Then God said, "Take your son, your only son, whom you love-- Isaac--and go to the region of Moriah. Sacrifice him there as a burnt offering on a mountain I will show you."

Early the next morning Abraham got up and loaded his donkey. He took with him two of his servants and his son Isaac. When he had cut enough wood for the burnt offering, he set out for the place God had told him about.

On the third day Abraham looked up and saw the place in the distance.

He said to his servants, "Stay here with the donkey while I and the boy go over there. We will worship and then we will come back to you."

Abraham took the wood for the burnt offering and placed it on his son Isaac, and he himself carried the fire and the knife. As the two of them went on together,

Figure 3.3 Sacrifice of Isaac

Isaac spoke up and said to his father Abraham, "Father?" "Yes, my son?" Abraham replied. "The fire and wood are here," Isaac said, "but where is the lamb for the burnt offering?"

Abraham answered, "God himself will provide the lamb for the burnt offering, my son." And the two of them went on together.

When they reached the place God had told him about, Abraham built an altar there and arranged the wood on it. He bound his son Isaac and laid him on the altar, on top of the wood.

Then he reached out his hand and took the knife to slay his son.

But the angel of the LORD called out to him from heaven, "Abraham! Abraham!" "Here I am," he replied.

"Do not lay a hand on the boy," he said. "Do not do anything to him. Now I know that you fear God, because you have not withheld from me your son, your only son."

Abraham looked up and there in a thicket he saw a ram caught by its horns. He went over and took the ram and sacrificed it as a burnt offering instead of his son.

So Abraham called that place The LORD Will Provide. And to this day it is said, "On the mountain of the LORD it will be provided."

The angel of the LORD called to Abraham from heaven a second time

and said, "I swear by myself, declares the LORD, that because you have done this and have not withheld your son, your only son,

I will surely bless you and make your descendants as numerous as the stars in the sky and as the sand on the seashore. Your descendants will take possession of the cities of their enemies,

and through your offspring all nations on earth will be blessed, because you have obeyed me."

Isaac's descendants were later called the Israelites. Abraham, Isaac, and his son Jacob are collectively venerated as the Patriarch of Judaism. The ten pre-flood or Antediluvian Patriarchs were Adam, Seth, Enos, Kenan, Mahalalel, Jared, Enoch, Methuselah, Lamech and the last one was Noah.

In the story of Isaac's birth a matter of surprise is that Sarah conceived Isaac at the age of ninety years while Abraham was a hundred years old. This was possible because God promised it to Abraham. I don't know about God's promise, but conceiving at such an old age of ninety is biologically rare but not impossible. It doesn't require any spiritual element. Even in modern times every now and then we get to see this kind of astonishing phenomenon of natural pregnancy over the age of 50 years. In 1887, The Lancet Medical Journal reported the case of an English woman who had given birth, at the age of 62 years ½, to three boys, real triplets, her 11th, 12th, and 13th children with her husband.

Later came Moses, one of the most important prophets of Judaism, as well as Christianity and Islam. Moses was a messiah to his people, the Israelites.

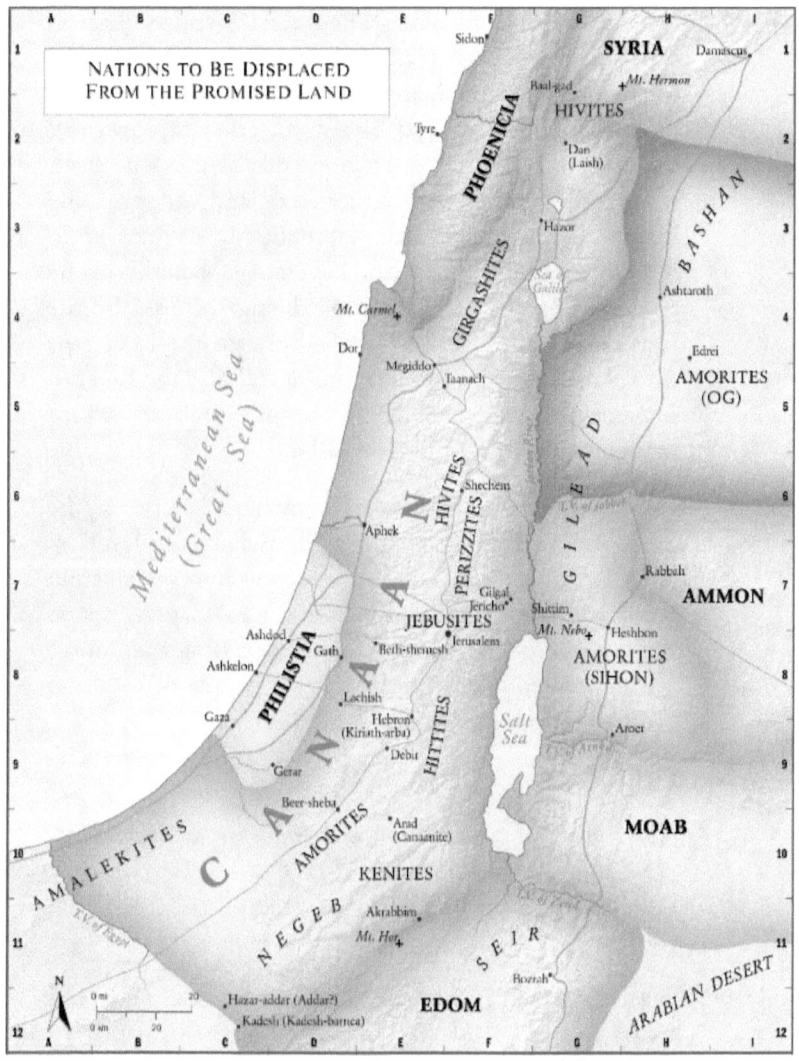

Figure 3.4 The Promised Land – CANAAN

He was born in a time when the Egyptian Pharaoh ordered all newborn Hebrew boys to be killed in order to reduce the population of the Israelites. Moses' mother Jochebed placed him in an ark and concealed the ark in the bulrushes by the bank of the Nile river, where the baby was discovered and adopted by Pharaoh's daughter Bithia. Naturally, Moses grew up in the royal family. But when he reached adulthood, he couldn't bear to see his people being tortured by the Egyptians any more. So he killed one of the Egyptians who was beating an Israelite and fled across the Red Sea to Midian. It was there at the Mount Horeb he had his first encounter with the God of Israelites.

This encounter with God was mankind's one of the most important steps through religious evolution. Moses vividly perceived God talking to him from within a burning bush. God commanded him to get back to Egypt and free the Israelites from their bondage. Moses was appointed to lead those chosen people of God out of Egypt into the Promised Land (Canaan).

Bible : Exodus 3 : 1-10

Now Moses was tending the flock of Jethro his father-in-law, the priest of Midian, and he led the flock to the far side of the wilderness and came to Horeb, the mountain of God.

There the angel of the LORD appeared to him in flames of fire from within a bush. Moses saw that though the bush was on fire it did not burn up.

So Moses thought, "I will go over and see this strange sight--why the bush does not burn up."

When the LORD saw that he had gone over to look, God called to him from within the bush, "Moses! Moses!" And Moses said, "Here I am."

Figure 3.5 The Burning Bush & Moses

"Do not come any closer," God said. "Take off your sandals, for the place where you are standing is holy ground."

The LORD said, "I have indeed seen the misery of my people in Egypt. I have heard them crying out because of their slave drivers, and I am concerned about their suffering.

So I have come down to rescue them from the hand of the Egyptians and to bring them up out of that land into a good and spacious land, a land flowing with milk and honey--the home of the Canaanites, Hittites, Amorites, Perizzites, Hivites and Jebusites.

And now the cry of the Israelites has reached me, and I have seen the way the Egyptians are oppressing them.

So now, go. I am sending you to Pharaoh to bring my people the Israelites out of Egypt."

Moses led the Israelites to Mount Sinai, where he had another cognitive encounter with God in which he received the Ten Commandments. Moses was there with the LORD forty days and forty nights without eating bread or drinking water. And he chiseled down the words of the covenant on two stone tablets.

The Ten Commandments go thus

I am the LORD thy God

Thou shalt have no other gods

No graven images or likenesses

Not take the LORD's name in vain

Remember the sabbath day

Honor thy father and thy mother

Thou shalt not kill

Thou shalt not commit adultery

Thou shalt not steal

Thou shalt not bear false witness against thy neighbor

Thou shalt not covet

We'll discover the Ten Commandments' real scientific origin in the next chapter. You'd be fascinated to find out that all the God-encounters of the patriarchs including the receiving of the Ten Commandments had no God element after all.

That is why alongside being beneficial to the humanity philosophically, some parts of the Hebrew Bible and all other scriptures are so damn absurd, outrageous and inhuman. Alongside true philosophical teachings like *"Thou shalt love thy neighbor as thyself"* from Mark 12:31 that inspired great men like Tolstoy, Gandhi and Martin Luther King Jr., we also see clear symptoms of megalomania, imperialism, superciliousness, short-temper, jealousy, sexism, bigotry, savagery and prejudice in the scriptures, for instance what we see in the Book of Deuteronomy.

Deuteronomy 31 : 1-2

"I am now a hundred and twenty years old and I am no longer able to lead you. The LORD has said to me, 'You shall not cross the Jordan.'

The LORD your God himself will cross over ahead of you. He will destroy these nations before you, and you will take possession of their land. Joshua also will cross over ahead of you, as the LORD said.

31 : 12

Assemble the people--men, women and children, and the foreigners residing in your towns--so they can listen and learn to fear the LORD your God and follow carefully all the words of this law.

31 : 13

Their children, who do not know this law, must hear it and learn to fear the LORD your God as long as you live in the land you are crossing the Jordan to possess."

32 : 12

The LORD alone led him; no foreign god was with him.

32 : 16

They made him jealous with their foreign gods and angered him with their detestable idols.

32 : 21

They made me jealous by what is no god and angered me with their worthless idols. I will make them envious by those who are not a people; I will make them angry by a nation that has no understanding.

32 : 25

In the street the sword will make them childless; in their homes terror will reign. The young men and young women will perish, the infants and those with gray hair.

32 : 39

"See now that I myself am he! There is no god besides me. I put to death and I bring to life, I have wounded and I will heal, and no one can deliver out of my hand.

32 : 41

If I sharpen my flashing sword and my hand grasps it in judgment, I will take vengeance on my adversaries and repay those who hate me.

32 : 42

I will make my arrows drunk with blood, while my sword devours flesh: the blood of the slain and the captives, the heads of the enemy leaders."

32 : 43

Rejoice, you nations, with his people, for he will avenge the blood of his servants; he will take vengeance on his enemies and make atonement for his land and people.

32 : 46-47

He (Moses) said to them, "Take to heart all the words I have solemnly declared to you this day, so that you may command your children to obey carefully all the words of this law. They are not just idle words for you--they are your life. By them you will live long in the land you are crossing the Jordan to possess."

These words of God through Moses are clearly filled with petty human characteristics. Literally there is not a single bit of spiritual element in these words from the Book of Deuteronomy.

Likewise, the Hindu epic Mahabharata, is filled with violence, savagery, ancient politics, sexism, imperialism and polygamy. For example, Lord Krishna already knew about the horrible fate of his nephew Abhimanyu in the war of Kurukshetra, yet he drove Arjuna (father of Abhimanyu) away from the battlefield and let Abhimanyu die brutally.

Figure 3.6 Abhimanyu fighting a bloody battle in Kurukshetra

And the most vibrant specimen of polygamy and perversion was Lord Krishna himself. He had mainly eight wives, for which he is also known as Ashtabharya (Ashta : Eight, Bharya : Wife). According to Bhagavata purana names of the eight wives in order were Rukmini, Satyabhama, Jambavati, Kalindi, Mitravinda, Nagnajiti, Bhadra and Lakshmana. Apart from those eight wives, he had 16,000 (or 16,100 in different texts) wives whom he rescued from the demon Naraka.

Naraka was the son of the earth goddess, hence he was also called Bhumasura (Bhumi : Earth, Asura : Demon). He had kidnapped 16,000 women and held them captive in his capital Pragjyotisha.

Narakasura had the boon of being killed only by his own mother. And Krishna's second wife Sathyabhama was considered to be the incarnation of mother earth. So Krishna and his wife Satyabhama attacked

Pragjyotisha riding on Garuda, the humanoid eagle. They fought a wonderful battle. Satyabhama killed Naraka and finally Krishna released 16,000 women that Narakasura had in captivity. When Krishna asked them to return to their houses, they refused. They were aware that the society of that age would not take back those who were taken by another man. So they were left with nowhere to go. When Krishna asked them what they wanted to do, they all wanted Krishna to marry them.

So Krishna married with all those 16,000 women making them his official wives so that they could live in honor. He also constructed them each huge palaces with enormous gardens and hundreds of maids. He also somehow managed to divide himself into 16,000 forms and engaged in proper marital life with all those wives.

Come on guys, after all, the author of Bhagavad Purana was an ancient sage Vyasa, a bronze-age human being just like the patriarchs of Judaism, who contributed to the Hebrew Bible. You can't really expect more from them. Consequently, their texts consist of good philosophical teachings as well as meaningless stuff.

The encounters with God didn't really end with Abraham, Moses and Vyasa. In fact, encounters with God increased over time.

Sometime around the 500-400 BC, Siddhartha Gautama experienced heightened state of spiritual enlightenment at the age of 35 after a presumed 49 days of meditation. He found the absolute peace and truth he was looking for. Since that time, Gautama was recognized by his followers as Buddha or "the awakened one". Now he had finally found the answer to all the sufferings of mankind: *"The cause of suffering is greed, selfishness and stupidity. If people get rid of these negative emotions, they will be happy."* Henceforth arose Buddhism.

Figure 3.7 The Giant 25 meters tall Buddha Statue in Bodhgaya, India

Then came the major Islamic scripture Quran. The Islamic faith of Quran is based on the spiritual experiences of Muhammad. He is considered to be the last of the prophets sent by God. The Muslims believe in oneness of God, just like the Advaita or non-duality concept of the Vedas. Muslims believe that after Adam, Noah, Abraham, Moses and Jesus, God finally sent the last prophet and messenger Muhammad to deliver the spiritual teachings. He is perceived to have restored

the unaltered origin   of   monotheistic faith
of Adam, Abraham, Moses and Jesus in Islam.

Figure 3.8 Archangel Gabriel teaching Muhammad

Muhammad was born approximately in the 570 CE in
the Arabian city of Mecca. Being orphaned at an early
age, he was raised under the care of his paternal
uncle Abu Talib. He primarily worked as a
merchant. Occasionally he would retreat to a cave in the
mountain Jabal al-Nour, for several nights of solitude
and prayer. It was there in a cave called Hira, on the
mountain Jabal al-Nour at the age of forty he reported
his first encounter with Archangel Gabriel.

In this encounter Archangel Gabriel apparently
commanded Muhammad to recite the first verse for
Quran. It was later recorded in the Quran as the first
Surah revealed to Muhammad, the Surah al-Alaq
(chapter 96).

Quran : Surah 96 (Al-Alaq), Ayat 1-5

Iqra/ bi-ismi rabbika allathee khalaq

Khalaqa al-insana min AAalaq

Iqra/ warabbuka al-akram

Allathee AAallama bilqalam

AAallama al-insana ma lamyaAAlam

Translation

Recite in the name of your Lord who created –

Created man from a clinging substance.

Recite, and your Lord is the most Generous –

Who taught by the pen –

Taught man that which he knew not.

Three years after this event Muhammad started preaching these revelations of Quran to people, declaring the oneness of God, just like Nanak's central teaching of Sikhism, "Ik Onkar" or "One God".

Sikhism was founded in the 15th century in Punjab, India by Guru Nanak as a monotheist religion. Again, like all other previous religions Sikhism was formed on the foundation of Nanak's divine experience with God. His story of encounter with the Lord goes like this.

Nanak used to take a bath in the river every day before sunrise. But one day he mysteriously disappeared with his clothes still lying on the river bank. Three days later he emerged from the water uttering *"there is neither Hindu nor Muslim"*.

Figure 3.9 Harmandir Sahib, also referred to as the "Golden Temple", the holiest Sikh Gurdwara in Punjab, India

Later he described his mystical encounter with God. He was taken to the court of God and given a cup of amrit (divine nectar) to drink. Then God commanded him *"I am with you, Go and repeat My Name, and teach others to do the same"*. He was so much filled with divine bliss that he composed his first foundational verse, known as the Mul Mantar (Root Mantra) for the central scripture of Sikhism, Guru Granth Sahib, which is considered to be the eleventh Guru. The entire Sikh faith is based on Mul Mantar :

Ik Onkar, Sat Nam, Karta Purakh, Nir Bhau, Nir Vair, Akal Murat, Ajuni, Saibhang, Gur Prasaad

Translation

There is only one God,

Eternal truth is His name,

He is the creator,

He is without fear,

He is without hate,

He is without form,

Beyond birth and death,

He is the enlightener,

He can be reached through the mercy and grace of the true Guru (teacher)

Nanak wanted to preach people that God loves both the Hindus and the Muslims the same way. Believing in his spiritual encounter, he wanted to eliminate the distance between the Hindus and the Muslims by teaching the words of equality and One God. But just like usual, he ended up forming yet another religion which became more and more hardcore with its own rituals and regulations in the hands of the subsequent nine Gurus, Guru Angad, Guru Amar Das, Guru Ram Das, Guru Arjan, Guru Har Gobind, Guru Har Rai, Guru Har Krishan, Guru Tegh Bahadur and Guru Gobind Singh.

History of human civilization is filled with these countless leaders, teachers, prophets and spiritual giants. Mahavira, Joseph Smith and many more. The list never ends. All had their own stories to preach. All had their own spiritual encounters to make others believe in. All of them thought that they had received the words of God.

But what really happened back then!!!

Sit back and enjoy the ride...

* * *

CHAPTER 4
HUMAN BRAIN CREATES GOD

"In any field, find the strangest thing and then explore it."

- John Archibald Wheeler

Here comes the strangest fact of neuroscience. It might hurt a little to the religious vanity of the human heart.

All those Gods and Angels that all the spiritual giants of mankind history have encountered were the creation of their brain and the background storyline of those experiences were guided by their own instincts and desires at their subconscious mind.

Human brain is an amazing apparatus that allows us to contemplate the universe, our own existence and even contemplate ourselves contemplating. But these wonderful abilities of the brain didn't occur overnight. Let's look at the figures. It took approximately six million years for the Homo sapiens brain to fully develop the marvelous characteristics of comprehension and logicality. At that point of time, six million years ago our ancestral line diverged from the line which developed into modern chimpanzees and other apes. Before that we were all in the same ape family. So the next time you see a monkey or chimpanzee, have some respect and may be invite him in for a hot beverage. Chimpanzees are our closest living relatives.

Throughout the evolution of mankind our very much primordial ancestors had one thing in common, it was ignorance. This ignorance gave birth to fear. Fear of the unknown became a quintessential element of their daily survival. To ace the intensity of the fear, rituals of worship arose.

Our earliest hominid ancestors didn't have ample brain capacity because it was too small in size. The ancient days were full of challenges. And as the early humans continued facing various environmental challenges they eventually evolved bigger body along with a larger and more complex brain. The challenges of surviving in that savage environment triggered the evolution of the brain size. Over the course of human evolution, brain size tripled. The modern human brain is the most complex of any living primate. Studies in genetics show that primates diverged from other mammals approximately 85 million years ago.

There were various factors that contributed to the development of the human brain. The climate compelled our early hominid ancestors to move from the lush forest to the savannah. We know this from the faunal remnants beside the early hominid fossils. This was the major contributor to the conditions evoking a kind of practical intelligence and emotional control.

The natural environment offered tools for primitive technological advancement and enabled them to begin the scientific journey. It was by the use of tools that our hominid ancestors began the slow process of forging a separation from the harsh and dangerous environment. The environment also selected the necessary mutations that provided us the biological means of survival. Organisms do not just adapt to an abstract, independent environment, they respond to their particular niche. This niche is often worked over thoroughly by the species that dwell in it.

We the Homo sapiens species belong to the Hominid family. In this taxonomic family our relatives are chimpanzees, gorillas and orangutans. Hominids diverged from the Gibbon family around 15 million years ago. Our most well-known early hominid forefathers were Ardipithecus and Australopithecus. Ardipithecus kadabba was one of the well-known earliest hominids who lived between around 5.8 and 5.2 million years ago.

However, the oldest human ancestors we know of are Sahelanthropus tchadensis and Orrorin tugenensis. They lived around 7 and 6 million years ago. But since the scientific community is yet to know more about them, we shall start from the species that evolved after them, i.e. Ardipithecus kadabba.

In the year 1997 paleoanthropologist Yohannes Haile-Selassie found a piece of lower jaw in the Middle Awash region of Ethiopia. At that moment he didn't realize that he had discovered a new species. But 11 specimens found later by 5 separate individuals convinced him that he indeed had found an early human ancestor. The specimens included hand and foot bones, partial arm bones, a clavicle and a toe bone.

Later in 2002, six teeth were discovered in the Middle Awash at the site Asa Koma. Based on these teeth, paleoanthropologists Yohannes Haile-Selassie, Gen Suwa and Tim White allocated the fossils in 2004 to a new species they named Ardipithecus kadabba ('kadabba' means 'oldest ancestor' in the Afar language). Ardipithecus kadabba was bipedal (walked upright), most probably similar in body and brain size to a modern chimpanzee. From Ardipithecus kadabba descended Ardipithecus ramidus.

One of the most important discoveries of early hominid ancestors was 'Ardi', a female Ardipithecus ramidus. She lived around 4.4 million years ago. American paleoanthropologist Tim White discovered the first Ardipithecus ramidus fossils in the Middle Awash area of Ethiopia in 1994. White's team has uncovered over 100 fossil specimens since then. He devised the genus name Ardipithecus to distinguish this new genus from the previously discovered Australopithecus.

In 2009, scientists formally announced the discovery. White nicknamed the 3 ft 11 inch (120 cm) tall partial skeleton "Ardi" (means "ground" or "floor").

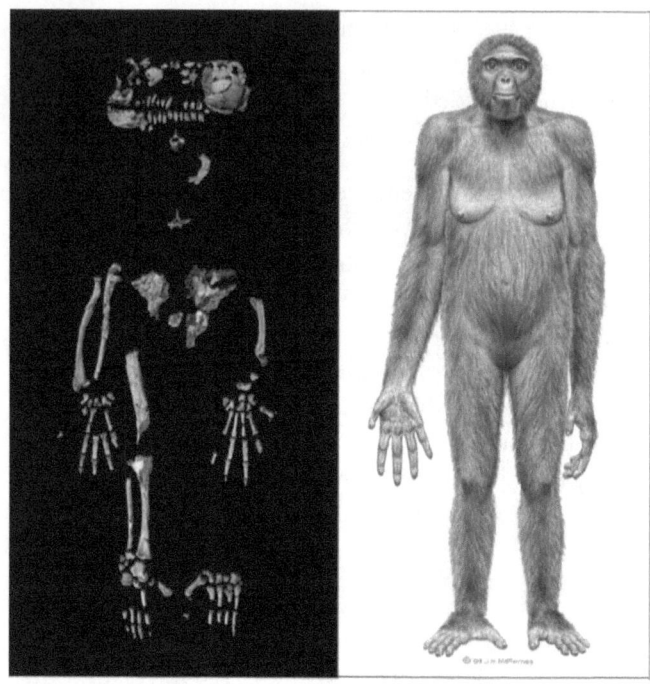

Figure 4.1 Left: Ardi's skeleton fragments, Right: Paleoartist Jay Matternes' rendition of Ardi

Ardi weighed around 110 lbs. (50 kg). Her foot bones indicate a divergent large toe combined with a rigid foot. The pelvis, reconstructed from a crushed specimen shows adaptations that combine tree-climbing and bipedal activity. Based on the fact that the size of the upper canine teeth in males is not much larger than the canines of females, we can say that Ardipithecus ramidus did not have much body size sexual dimorphism, which means a male individual would have been similar to a female in size. It's possible that the males did not compete against each other for dominance and therefore did not need to grow bigger in size. Ardipithecus ramidus had a small brain, measuring between 300 and 350 cc. They required to climb tree grasping with big toe. For this they needed to develop more accurate motor skills. So the brain size increased to around 450 cc in the later genus of hominids, Australopithecus.

Australopithecus played a significant part in human evolution. The genus Homo derived from Australopithecus at some time after 3 million years ago. In the roughly 2 million years of existence, Australopithcus branched into at least five different species.

Australopithecus afarensis was the first among them and one of the longest-lived and best-known early human species. The first Australopithecus afarensis skeleton was discovered on November 24, 1974 near Hadar in Ethiopia by Tom Gray and Donald Johanson. She was named as "Lucy". The next year Donald Johanson's team discovered another site in Hadar which included over 200 fossil specimens from at least 13 individuals, both adults and juveniles. These fossil specimens are commonly referred to as the "First Family".

Lucy and the First Family were very important discoveries of human evolution. That's why this species is also called Lucy's species. They lived in Eastern Africa (Ethiopia, Kenya, Tanzania) between around 3.85 and 2.95 million years ago.

Figure 4.2 Left: Lucy's skeleton fragments, Right: Reconstruction of Lucy

Lucy's species survived for more than 900,000 years, which is over four times as long as our own species has been around. They had both ape and human characteristics such as apelike face proportions (a flat nose, a strongly projecting lower jaw) and braincase with a small brain, usually less than 500 cc and long, strong arms with curved fingers adapted for climbing trees. They also had small canine teeth like all other early humans, and a body that stood on two legs and regularly walked upright. Their adaptations for living both in the trees and on the ground helped them survive for almost a million years as climate and environments changed.

Then there lived Australopithecus africanus, our direct ancestor. The first skull of this species was unearthed in 1924, in the lime mine at Taung near Kimberley, South Africa. This skull was nicknamed "Taung Child". Raymond Dart was sure that it was a new intermediary species between ape and humans and named it Australopithecus africanus which means 'southern ape of Africa'. But it took more than 20 years for the scientific community to widely accept it as a member of the human family. They lived between about 3.3 and 2.1 million years ago.

Australopithecus africanus with a brain size of 500 cc, contributed directly to our hereditary line. The pelvis, femur and foot bones indicate that they walked bipedally, but the shoulder and hand bones indicate they were also adapted for climbing. The fossil remains of Au. africanus show that they eventually left the trees for life on the ground except when chased back sporadically by the big cats which dominated the area. Because of their massive jaws and teeth, they were believed to have had a diet similar to modern chimpanzees, which consisted of fruits, plants, nuts, seeds, fibrous roots, insects and eggs.

No stone tools have been discovered in the sediments alongside Au. africanus fossils. They were not good hunters. Broken bones of carnivorous animals like lions, leopards and hyenas were found alongside their fossils. These predators even ate Au. africanus individuals. Their average life span was 30 years but children and females were particularly vulnerable to many larger carnivores. So naturally it was not at all a safe environment for survival.

Eventually they had to leave the forest altogether and move to the savanna where their upright posture helped to see longer distances for scavenging food and watching for predators. Slowly hominid legs became longer. They developed arches in their feet allowing them to cover more ground than many of their four-legged cohabitants.

Other Australopithecine species were Au. anamensis, Au. garhi and Au. sediba.

Early Australopithecines lived in groups a little larger than modern chimps but this size increased with later hominid species. They had not yet developed advanced stone tools so their only survival kit was their social groups. These social groups provided the foundation of an emotional brain. Our emotional instincts are bequeathed to us by these ancestors. Our Australopithecine ancestors' communication was emotional. But on the savanna, negative emotional outbreaks could disrupt the group as well as make noises that would attract predators. This created adaptive pressure both for cortical control of emotion and for the so called basic social emotions of sympathy, guilt, and shame which made way for the increase in the brain size mostly in the neocortex that added an extra layer to the whole brain and made room for more neurons. More neurons mean more cerebral activity. Whatever social intelligence the Australopithecines possessed did not spill over to tool use and hunting strategies. Communication was confined to physical gestures and vocalization.

It was the Homo habilis who made the first stone tools. Au. garhi were also capable of making primitive tools but they were not so sophisticated as the habilis' tools. A team led by scientists Louis and Mary Leakey uncovered the fossilized remnants of a unique early hominid between 1960 and 1963 at Olduvai Gorge in Tanzania. The specimen found by Jonathan Leakey was nicknamed "Jonny's child".

Figure 4.3 Reconstruction of Homo habilis

This early human had a combination of features different from those seen in Australopithecus. Louis Leakey, South African scientist Philip Tobias, and

British scientist John Napier declared these fossils a new species and called them Homo habilis (meaning 'handy man'), because they suspected that it was this slightly larger-brained early human that made the thousands of stone tools that were found at Olduvai Gorge.

Homo habilis lived between 2.4 and 1.4 million years ago in Eastern and Southern Africa. As they learnt the use of tools, they emphasized on including meat in their diet. This led to the significant augmentation in their brain size. The increase in the brain size relative to the body size is called "encephalization". The pattern of significant encephalization began with Homo habilis. Brain capacity pumped up to about 550 cc some 100 cc more than Australopithecines. This increased to 800 cc, toward the end of their existence. The encephalization included the development of the language center the Broca's area in the frontal lobe.

Homo habilis existed for 1 million years, but despite the initial advancement in stone chopper tools the species did not go on making further refinements in efficiency in their lifetime. They never ranged outside of Africa and we find no indication of language capacity despite the development of the language center in the brain. Culture must have been confined to the moment with little concern for the past or future. Aside from their first use of stone tools, they neither innovated nor explored.

Things picked up speed with the arrival of Homo erectus around 1.8 million years ago. During their existence they excelled in their skill of making stone edges and the use of fire emerged. Their greatest advantage was their immensely increased brain size. Their cranial capacity doubled from 550 cc to 1,100 cc along with increased complexity of the neural network.

Homo erectus had a wider inventory of tools than earlier hominids and their communicative capacities and general sociality greatly increased. The front of the head expanded and the face flattened to accept the increase in the frontal, temporal, and parietal lobes.

Figure 4.4 Reconstruction of a female Homo erectus by John Gurche

The expansion in the temporal lobes is what we needed to experience the vivid encounters with God and to experience spirituality.

Cognitive functions started to develop beginning with imitation and mimicry involving vocalizations, facial expression, eye movements and emotional expression. This was an important phase for the development of a truly social brain. At this stage of encephalization the limbic brain got totally rewired via the newly developed cortex. This gave rise to the social emotions of shame and guilt that taught those ancestors of ours to be cautious about their social behavior. This significantly implies to the dawn of a "self" in those early human ancestors. With their newly developed cranial capacity they started to migrate out of Africa to the southern Asia and Europe.

Today's famous concept of two separate brain hemispheres having separate characteristics arose from the brain of the Homo erectus. Their brain was lateralized to create two different hemispheres. We'll illustrate the qualities of the left and right hemispheres of the brain in a little while.

Over time Homo erectus learnt to make their tools more refined, symmetrical and sharper. Their inventory of tools increased to include hand axes, cleavers and knives. This gave them the capacity to free themselves from the dictates of the harsh environment and survive in the harsher climates to which they traveled. They used water to migrate to their summer homes in southern France, even to Southeast Asia and to return to North Africa in the winter. With their increased brain functions they were able to create mental maps of the regions they covered. With all these strengths Homo erectus had an enormous gift for withstanding monotony. After developing their expertise in making tools they lived for more than one million years without further development.

Then came the Homo heidelbergensis who lived between 700,000 and 200,000 years ago. This was the first species to build shelters out of wood and rocks. From them descended two hominid species, the Homo neanderthalensis and Homo sapiens.

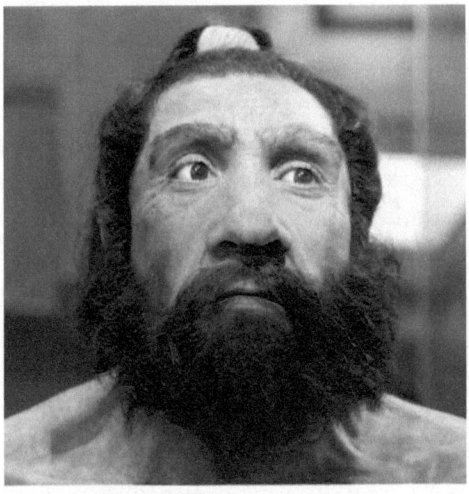

Figure 4.5 Reconstruction of Neanderthal man by John Gurche

Homo neanderthalensis are known as the Neanderthals and they were our closest human relatives. Their brain size bumped up to around 1,500 cc, which was a little bigger than our sapiens brain. Their body was more robust with thicker bones and more muscle. They introduced the first composite tools that were made from at least three different materials conducive to hunting big game. Their skill at making sharp stone edges was remarkable and like Homo erectus, they somehow achieved the feat of passing this capacity on through generations despite the absence of speech.

The sense of aesthetics was first seen in the Neanderthals. Archeological evidences eloquently illustrate that it was the Neanderthals who called for the dawn of religious belief. Burial sites discovered throughout Eurasia imply their belief in afterlife. They consciously buried their dead and occasionally even marked their graves with offerings, such as flowers. They made and used a diverse set of sophisticated tools, controlled fire, lived in shelters, made and wore clothing, were skilled hunters of large animals and occasionally made symbolic or ornamental objects. Despite all their advancements Neanderthals existed only until 40,000 years ago.

We, the Homo sapiens evolved around 200,000 years ago in Africa from Homo heidelbergensis. Within 50,000 years we inhabited all corners of the planet earth. The brain size reached its present size of an average 1,300 cc, with a fully expanded frontal lobe and other cortex regions. The larger left hemisphere allowed the true development of speech production with the use of the language centers of the brain, Broca's area (in the frontal lobes), the region which houses the capacity to produce grammatical speech and Wernicke's area (behind the temporal lobes) which makes possible the semantic understanding of words and the reception of speech. Naturally, unlike any other previous hominid species, we began to talk using not only vocabulary but proper grammatical syntax. Homo sapiens had proper vascularization throughout the entire cortex that allowed proper blood supply for various complicated brain functions.

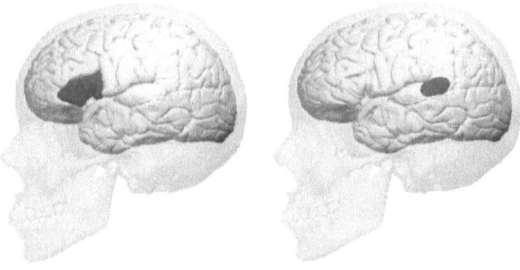

Figure 4.6 Left: Broca's Area, Right: Wernicke's Area

Then occurred what we scientists call the Big Bang of Evolution around 90,000 years ago. Homo sapiens started a revolutionary way of living. Technology increased at a great pace. Our brain's neural circuitry became way more complicated and the interaction between the synapses became more intricate than our early hominid ancestors. And all that the prophets needed to encounter God was a nicely developed cerebral cortex deeply interconnected with the limbic system underneath. More neural connections mean more complex neural firings. This gives rise to the risk of misfiring or overfiring within the neural network of the brain. And whenever this kind of abnormal neural firings occur in specific regions of the brain they create exotic experiences such as encounter with God. It is what we neuroscientists call "Parasitic Consciousness", from which the book got its name "The God Parasite".

Experiencing God, angels or ghosts is just one among many vivid 'ictal' experiences of the Parasitic Consciousness. These vivid ictal experiences include sensing the presence of another being ('sensed presence'), out-of-body experiences, tunnel vision etc. All these experiences are purely connected to anomalous activity of various parts of the brain. For example, most of the exotic encounters with spiritual beings are neurologically associated with anomalous temporal lobe functions. The anomaly can occur naturally within the brain, or due to tiny lesions or by the influence of disturbance in the geomagnetic field of our planet.

During the whole encephalization process, Homo sapiens and their early hominid ancestors' brains have been immersed in the geomagnetic field of mother earth. With the expansion of a complex cerebral cortex, the human brain became more susceptible to the disturbances occurring in the geomagnetic field of planet earth. As a result, any slight disturbance to the geomagnetic state of our beloved mother earth can evoke various mystical experiences within our brain.

Modern neuroscience evidently shows that all experiences are generated within the brain. It is a fundamental principle of cognitive & behavioral neuroscience. There are several important factors to this principle. All experiences are responses that are purely evoked by physical event or stimuli.

For example, individuals who survive a traumatic life-threatening event often report experiencing out-of-body experiences. They describe that during the traumatic episode their 'self' actually left their body and visited a spirit world which they often interpret as heaven, where they met their dead relatives. Here the intense stress of the deadly trauma releases increased amount of stress hormones like cortisol and adrenalin. This high level of stress hormones induces the release of DMT (Dimethyltryptamine) in the pineal gland. DMT is a profoundly hallucinogenic compound which creates the virtual reality of the spirit world with the help of the right temporoparietal junction. Here the utter stress acts as a physical stimuli that evokes the paranormal experiences during a traumatic episode. Although the purpose of increased production of DMT in the pineal gland is to assist in healing from the trauma, it gives the traumatized individual an experience of a lifetime in the process.

Neurologist Olaf Blanke of Geneva University Hospital in Switzerland demonstrated the creation of out-of-body experiences in volunteers by stimulating the right temporoparietal junction of their brain.

Neuroscientific literature illustrates that all human experiences, from the feeling of love, to the presence of God emerge from various brain activity. And by isolating the controlling stimuli that evokes a mystical experience we are able to create that exact experience inside the laboratory. So, even the experience of God is subjected to experimental verification.

Thus, we actually have the answer to how exactly the spiritual leaders of mankind history had encounters with

God and angels and experienced spiritual transcendence. The answer lies in your temporal lobe. Recorded history of neurological evidence shows that vivid encounters with spiritual beings are in fact evoked by transient, electrical microseizures within the temporal lobes of the human brain, especially the right temporal lobe. To the layman, epileptic seizures mean someone having powerful involuntary contraction of all muscles of the body and losing consciousness. Indeed, these symptoms characterize the most well-known form of epilepsy, called a grand mal seizure. Such seizures usually arise because a tiny cluster of neurons somewhere in the brain starts misbehaving, and therefore fires chaotically until anomalous neural activity spreads like wildfire and engulfs the entire brain.

But seizures can also be "focal"; that is, they can remain confined largely to a single small patch of the brain. Such focal seizures (also known as partial seizures) can cause a kaleidoscope of symptoms depending on the location. For example, if such focal seizures take place in the motor cortex, the result is a sequential march of muscle twitching, what we call jacksonian seizures. When focal seizures take place in brain regions such as the temporal lobe, the individual experiences the sense of oneness, complex hallucinations and euphoria. Often, these seizures don't involve any convulsions at all and in some cases they can invoke the presence of God.

During these God experiences, whatever messages the individual apparently receives from that illusory Supreme Being are just his/her own gut instincts and subconscious desires. During these microseizures an individual who craves for philosophical guidance would receive the gospel from his own instincts through the imaginary spiritual being, likewise a person with nomadic instincts would vividly hallucinate the promise of a new land.

Nomads believing themselves to be "chosen" have marched into lands belonging to others and slaughtered them, all because of a vivid gut feeling *"God gave us the land"*. Anomalous temporal lobe activity led to the arrival of the Ten Commandments. Likewise, a person with insanely pervert subconscious desires would justify those polygamous desires as the command of God that eloquently happened to Joseph Smith, the founder of The Church of Jesus Christ of Latter Day Saints.

The most crucial beautiful literary contribution to this mystical phenomenon was done by the Russian Novelist Fyodor Dostoyevsky. He kept records of his 102 epileptic seizures during his last two decades, which mainly occurred at night. Seizures which occurred in the daytime were often preceded by an ecstatic aura. This has led neurologists to theorize that he had temporal lobe epilepsy. Based on his experiences, he created characters with epilepsy in his four novels The Possessed, The Brothers Karamazov, The Insulted and Injured, and The Idiot.

In "The Idiot", he describes such "religious seizures" through the character of Prince Myshkin:

"He [Myshkin] remembered that during his epileptic fits, or rather immediately preceding them, he had always experienced a

moment or two when his whole heart and mind, and body seemed to wake up to vigour and light; when he became filled with joy and hope, and all his anxieties seemed to be swept away for ever; these moments were but presentiments, as it were, of the one final second (it was never more than a second) in which the fit came upon him."

Dostoyevsky also recorded his own seizure experiences as:

"For several instants I experience a happiness that is impossible in an ordinary state, and of which other people have no conception. I feel full harmony in myself and in the whole world, and the feeling is so strong and sweet that for a few seconds of such bliss one could give up ten years of life, perhaps all of life.

I felt that heaven descended to earth and swallowed me. I really attained god and was imbued with him. All of you healthy people don't even suspect what happiness is, that happiness that we epileptics experience for a second before an attack."

His writings are a treasure chest of blissful epileptic experiences that clearly associates the experience of God with the most profound feelings of human existence.

Many modern day temporal epileptics have reported of a "divine light that illuminates all things," and of an "ultimate truth that lies completely beyond the reach of ordinary minds who are too immersed in the hustle and bustle of daily life to notice the beauty and grandeur of it all."

Now comes the question, how do various people experience God in different forms and figures?

Well, that's even more fascinating.

The most conspicuous feature of the human brain is the two hemispheres. The total surface area is about 1570 cm². The right hemisphere contains about 40 cm² more area than the left one. While the left hemisphere is involved with your sense of self, the right one is responsible for your sense of the 'other'. This is a kind of awareness that keeps your mind vigilant about every single element around you but yourself.

And the fundamental origin of all kinds of spiritual or sentient presence is in the right hemisphere. This side of your brain is hardwired to see the Big Picture, even when there is no actual picture. The brain draws the storyline of the spiritual/religious experience based on the person's fantasies, attitude, desires and beliefs. This way God takes various forms and figures through the illustrations of various interpretations.

In the year 1838 French psychiatrist Jean-Étienne Dominique Esquirol first recognized that temporal lobe epilepsy was the root cause of all kinds of spiritual/religious experiences. A specific experience during an epileptic fit was reported in 1872. The patient believed that he was in heaven. He would appear to have been depersonalized, as it took three days for his body to be reunited with his soul according to the patient. He expressed that he was now a new man and had never before known what true peace was.

Often temporal lobe epileptics report that God has given them a mission to transform the whole world, just like Joan of Arc claimed that she had visions of the

Archangel Michael, Saint Margaret and Saint Catherine instructing her to support Charles VII and recover France from English domination. We have definite proof that she suffered from tuberculosis and with a temporal lobe tuberculoma. A tiny brain lesion in her temporal lobe triggered her vivid visual and auditory hallucinations of various angels.

Joan of Arc was born in Domremy, 1412. Her father has variously been described as a ploughman, keeper of a cattle pound, and a low-ranking official. Her early childhood, in the village where she was born, has been described as harsh and she received no formal teaching though her mother taught her basic household skills. She also tended her father's cattle, though legend has it that she was a shepherdess. She was about 13 when she first heard voices. The description of this first experience is quoted in Smith's Joan of Arc:

"She had a voice from God to help her to know what to do. And on this first occasion she was very much afraid ... She heard the voice upon the right side and rarely heard it without accompanying brightness ... after she heard this voice upon three occasions, she understood that it was the voice of an angel".

She later went on to claim that she heard and saw St Michael, St Catherine and St Margaret. Her behavior in her teenage years was quite exceptional for a female living during that period. She raised a siege, won a battle and tried to help the Dauphin of France regain lost territory from the English and Burgundians. Her character has been gracefully described as pious, brave, charming and lovable but also as quite ruthless and cruel. She was also well known for her avowed virginity and her liking to dress in male clothes.

Eventually she engaged in a battle with the Burgundians even though she knew she was outnumbered by them and was finally captured. On this occasion she claimed that she had been misguided by her voices. The Burgundians handed her over to the English for a sum of money in 1431 and she stood trial. She was found guilty but signed a form of abjuration and was condemned to imprisonment. However, later she was sentenced to be burnt at the stake. She died exactly in that manner and it was well documented that her heart and parts of her intestines did not burn and were later collected and thrown into the River Seine.

All these facts can be explained by the hypothesis that Joan of Arc suffered from tuberculosis with a temporal lobe tuberculoma and tuberculous pericarditis. Calcification of the tuberculous mesenteric glands and chronic tuberculous pericarditis could account for the heart and parts of the intestines being intact after she was burnt at the stake. A tuberculoma in the temporal lobe of her brain could account for her hallucinatory experiences.

So far in various neurological studies, experiences like déjà vu, time distortion, sensed presences of God, angels, ghosts or aliens have been elicited in healthy individuals by stimulating various regions of the brain, especially the right temporal cortex.

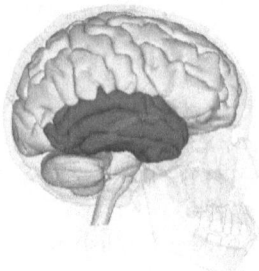 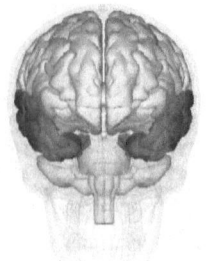

Figure 4.7 Left: Lateral view of Temporal lobe, Right: Anterior view

In the studies of Michael Persinger, cognitive & behavioral neuroscientist at Laurentian University the application of complex magnetic fields over various regions of the right hemisphere has consistently generated experiences of God or religious entities in people who believe in them. For example a 50 year old true Christian man reported that he suddenly felt and eventually visualized the presence of Jesus. Some volunteers reported the presence of bizarre beings, guardian angels, the devil and even aliens.

The science writer Michael Shermer also visited Persinger's lab to become one of his volunteers. He as well reported the experience of sensing something or someone passing next to him and his self trying to leave his body (mild out-of-body experience). The famous biologist and atheist Richard Dawkins went through the same experiment but didn't feel anything significant. In a personal communiqué Dr. Persinger explained to me how the person's own beliefs and fantasies influence his ability to interpret those experiences.

The temporal lobes are highly interconnected with the limbic system, especially the amygdala. Amygdala is basically our emotion and fear center. When the temporal lobe acts anomalously, it triggers excitement in the amygdala, which often leads to a feeling of unearthly bliss and absolute meaningfulness of the universe. Such feeling often overwhelms an individual during the God encounter. Everything around him/her is imbued with cosmic significance. Individuals often exclaim with utmost joy, *"I finally understand what it's all about. This is the moment I've been waiting for all my life. Suddenly it all makes sense."* or, *"finally I have insight into the true nature of the cosmos."*

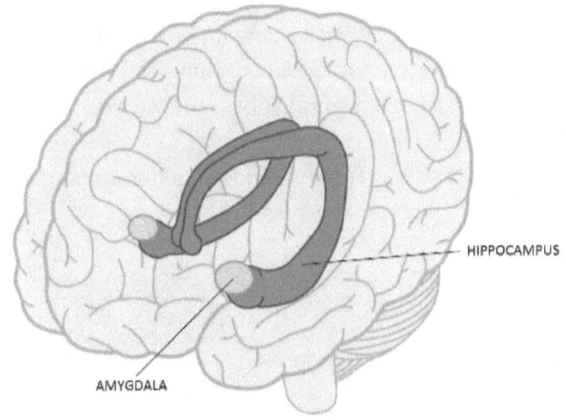

Figure 4.8 Amygdala and Hippocampus

The seizures last usually only for a few seconds each time. But these brief temporal lobe storms can sometimes permanently alter a person's personality. The repeated electrical bursts inside the patient's brain permanently facilitate certain pathways or may even open new channels of perception. This process, is called kindling and might permanently alter an individual's inner emotional life.

Specific electrical anomalies in the amygdala during the temporal lobe storms can also trigger aggression and homicidal urges. This counts for the outrageous and murderous commands from man's God. We humans are the Tyrannosaurus Rex of mammals.

Technology within the last two decades has been particularly revealing about the neurological foundation of religious/spiritual experiences. Functional magnetic resonance imaging (fMRI), positron emission tomography (PET), and single positron emission computer tomography (SPECT) have shown specific activity within the temporal lobes and adjacent parietal connections during sensed presence and imposed sensory deprivation. This expands our understanding of the meditative state of the Buddhist monks.

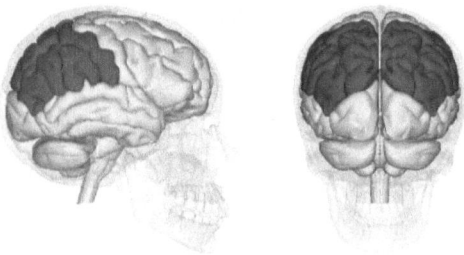

Figure 4.9 Left: Lateral view of Parietal lobe, Right: Posterior view

Andrew Newberg, a neuroscientist at the University of Pennsylvania used single photon emission computed tomography (SPECT) to examine the brain activity of Tibetan Buddhist monks as they meditated. Newberg found increased activity in the frontal lobe, which deals with concentration. This means that the monks were obviously concentrating on the activity.

Another significant result of his study was immense decrease of activity in the parietal lobe. The parietal lobe makes a person aware of his location in three-dimensional space. This lobe helps you look around to determine that you are sitting on your couch or chair about 10 feet away from the door, 6 feet away from the window and so on. The decreased parietal activity in the brains of the meditating monks indicates that they lose their ability to orient in three-dimensional space and they can no more differentiate where they end and something else begins. In this state of mind they become one with the universe, a state often described in the moment of transcendence or Samadhi.

Mankind history is filled with individuals around whom many religions have evolved. And modern neuroscientific literature strongly implies that all those individuals were overwhelmed with temporal lobe storms and in some cases decreased parietal lobe activity. The God experiences occurred in the mountains (Moses), in the caves (Muhammad), during walks in the wilderness (Christ), during imposed sensory deprivations (Buddha) and so on. Each of them was convinced that they had accessed the cosmic consciousness of the universe. As a result, their atypical cerebral activity inaugurated the modern religious movements that altered the course of history forever. These religious movements ultimately affected the development of the human civilization.

* * *

CHAPTER 5
GOD - HEALER OR KILLER

"We may know who we are or we may not. We may be Muslims, Jews or Christians but until our hearts become the mould for every heart we will see only our differences."

- Rumi

Humans do indeed need religion. Yes, but to what extent! Is it really worth relying on the books that were written thousands of years ago, for every insignificant problem we face? Those who wrote the so called scriptures didn't have technologies of today, so they had to rely on supernatural explanations of things. It is true that some parts of those scriptures have great philosophical potential. But as they were written by technologically primitive human beings, they have petty human characteristics as well which were endowed on their Gods.

We are now not primitive any more. We can't just blindly chant every single verse from the scriptures. Human beings are totally capable of distinguishing between good and evil, thanks to the pre-frontal cortex. We are not anymore controlled by our limbic brain which is irrationally emotional. Being modern humans we now have the ability to form a state of constructive equilibrium between our analytic pre-frontal cortex and the emotional limbic brain.

God is the placebo for the masses. In modern medicine for various psychological conditions medical practitioners often use sugar pills as a placebo to deceive the patient in believing that those pills are actual medicines which would treat the condition. The entire concept of placebo effect is based on fake treatment that often proves clinically effective. Likewise, even though the so far perceived omnipresent being doesn't really exist, sometimes just believing that it does exist and looks after people, gives them the will and strength to survive the daily struggle of life. This belief has evolved through millions of years, so it will be totally foolish and naïve to think that we can get rid of this illusion in a few decades or even centuries.

When a person holds a strong belief, no matter how irrational it is, you cannot really mend that person's perception with rationality. That's because the person grows a psychological immune system to defend the belief. Rationality cannot penetrate that shield so easily. And if somebody with utter influential potential finally succeeds taking away people's God and religion, that person will end up being their God himself. Because some individuals might be strong enough not to need God, but people are weak. God is the human brain's natural anti-depressant. Every time a trauma hits people, they cling to their God for survival. In time of distress 'hope' keeps them alive.

Beliefs are various cognitive structures that allow us to minimize anxiety of the unknown. And the most powerful sentence that shows the anxiety of the unknown at its height is *"I will die"*. These three words have maintained the God belief in various human cultures so far where most people have not directly encountered God.

Belief in God is just a rudimentary strategy of human civilization to remove the anticipation of termination. Every culture has devised its own word or concept that refers to some component of the infinite, omnipresent and omniscient. It is really simple this way. If something is infinite, there is no end. If there is no end, there is no anxiety. This semantic operation of the human mind required language structures that define the self as a subset of the infinite. Hence, phrases like *"I am the child of God"*, *"I will live on with God"* were born. Thus one can now easily say that God is eternal and one's self does not die because it becomes one with the eternal.

God heals as well as God kills. Uranium has the potential to light up a whole big city and even a country. Being said so, it also has the potential to incinerate a whole nation to ashes. God is indeed a delusion. There is no denying that. God is created by the brain and can also be destroyed by the brain. But this way 'Love' doesn't exist either, because it also is the result of molecular interaction within the brain. Love is a result of complex and mind-blowing interactions between neurotransmitters like Oxytocin, Dopamine, Serotonin, Endorphins and the reward center of the brain.

Does this mean we should downright stop loving our family or friends? Since there is no existence of actual 'Love', should we just deny what our heart feels for our beloved ones? Whatever the answer is, the same goes for the faith upon divine existence. Only one thing to remember – Love gives life, so should faith. As the neocortex of the brain keeps getting more complex through further evolution, eventually our far away progeny will be born in a world where there will be no more religion to be endowed upon them.

Last but not the least, let's have a little experiment which you can participate in right now. I'll simply suggest a few scientifically factual statements for you to either accept or defy. It will only require a basic idea about the human brain which I believe you already possess.

Now, listen carefully.

It's necessary that you read in the exact order they are given. And you can take not more than 5 seconds to decide at each statement. Do not skip any of the statements.

Get set….

Go

1. The neuron is a cell!

2. The brain is composed of neurons!

3. Different interactions between neurons produce different behavioral responses!

4. All perceptions occur within the brain!

5. All experiences occur within the brain!

6. Beliefs are composed of experiences!

7. Most people believe in the existence of God!

8. God is a belief!

9. So, God is an experience within the brain!

10. All experiences are produced by the brain as behavioral responses to various events!

11. The experience of God is produced by the brain!

12. If all the above statements are found to be true, then God does not exist except within the brain!

It doesn't really matter whether one is religious or an atheist. The simple fact of neuroscience is that, all kinds of beliefs (religious/non-religious/rational) are the product of various brain circuits and they serve only one purpose, i.e. maintain mental stability. Perhaps the easiest way to ensure such stability is to believe in some transcendent higher power that controls our destiny. No wonder temporal lobe epileptics experience a sense of omnipotence and grandeur, as if to say, *"I am the chosen one. It is my duty and privilege to transmit God's work to you lesser beings."* Regardless of the perpetual battle between believers and atheists, for me, religion is a tool for making friends, rather than making enemies.

*   *   *

BOOK IV

THE SPIRITUALITY ENGINE

INTRODUCTION

Why do we believe what we believe?

Why some of us are religious, some spiritual, some extremists, some agnostics and some atheists?

Why there is one elementary commonality among all religions, i.e. Spirituality?

What is it that makes various human beings perceive different things at different moral, logical and spiritual level?

Are we all merely smart biological devices with various operating platforms like iOS, Android, Windows, Linux and Blackberry?

What makes us so different in terms of beliefs and disbeliefs?

And the ultimate question is, does Spirituality have any divine element to it at all, or is it all very much biological? In that case, why do we even have this biological system of spirituality?

Does it have any significance in the human biology?

271

Spirituality is some magical twine that binds all the religions of the entire world together with the touch of blissful divinity. Even some people who don't consider themselves to be so much religious, they often proclaim to be highly spiritual. That's just a way of saying, *"I don't give a damn about which God rules the universe. I just prefer to have the divine peace within me as a secular humanist".*

Often the atheist part of the society claims that religion has killed more people throughout mankind history than wars. And to be honest, that is unfortunately true. The so called poppycock rules and rituals of religions often compel humans to fight with each other in a battle of proving one's religious superiority to the other. So, technically the atheistic point of view is correct about religion being the cause of countless homicides throughout history.

But, the common essence of all religions that is Spirituality functions within the human mind with or without the religious factors involved. Spirituality works as an engine within the human brain to inspire a human being in his/her all walks of life. Religious rituals and regulations are all manmade paths that are perceived to lead towards the Spiritual awakening. The mystical or spiritual experiences of the religious leaders of mankind history were the products of their atypical brain activity, as you might have read in my last book The God Parasite.

To quote from that book, *"The entire belief system on a Supreme Creator was created by primordial humans out of fear and ignorance, by means of illusions. Our primitive ancestors*

272

handed over their ideas of a supernatural sentient creator to us and as our brain size augmented over the period of evolution, we made the concept of God more sophisticated by means of our sophisticated hallucinations." Here these hallucinations of the spiritual giants were induced by either malfunction or over-excitement in specific regions of their brain. The blissful state they experienced are conceived to be the ultimate state of Spirituality.

Even though all the human perceptions of a Supreme Creator or God are based on fear, anxiety, illusions and neurological disorders, why are these beliefs still so crucial to humans that many are even prepared to kill or even die for them? The Gods we have perceived so far do not exist anywhere except within our head. The primordial humans feared any powerful manifestation of the nature. Hence they started to worship those natural elements only to reduce their anxiety and fear. That's Darwinism people. The human species still has religions because the majority of humans need them to survive the daily struggle of life. Human brain is designed to believe in the supernatural. Mother Nature embedded the seeds of spirituality within our genetic map because this bizarre quality proved its mettle throughout evolution.

But again, human brain is susceptible to new beliefs. Sounds confusing right! It simply means a believer can become an atheist as well as an atheist can become extremely religious. The same goes for spirituality. The brain creates Spirituality as well as undermines it as per the need of the moment. It's all about survival. Spirituality, Religiosity, Morality and Logicality all these human qualities are literally the babies of the brain that act as the means of human survival.

* * *

Chapter 1
The Spirit Emerges

In the permeating darkness of our distant ignorant and primitive past, our primitive ancestors had no modern means to satisfy their curious yet childishly immature mind, whenever they encountered something above their level of understanding. They were the slaves of Mother Nature. They were tormented in every walk of life by the surrounding harsh environment.

Nature didn't show any mercy to any species on earth. And humans were no exception either. Our poor ancestors had to fight for their survival against the nature that only put forward obstructions in their path. They had very little opportunity to develop their intellectual side. They feared every single natural manifestation of power, like thunderstorms, droughts, rain, floods, earthquakes, landslides, volcanoes, fire, heat, and cold etc. To avoid intellectual effort they had to incorporate supernatural explanations to those phenomena. They basically suspected that all those natural powerful events were Mother Nature's way of showing that she was angry with our ancestors. Naturally, our primitive ancestors felt the urge to appease that anger by worshipping nature.

275

Hence, they started to worship all the elements of Mother Nature like stones, hills, trees, lakes, animals and many more in the name of early spirituality. These were the early signs of spirituality that built the fertile ground for modern religions. Now the question emerges, what's really the difference between spirituality and religion?

Modern religions arrived much later when the humans had already learnt to tame the natural forces. The spiritual mind of our early ancestors led towards the urge to come up with religious rituals in order to sustain the spirituality throughout generations.

Modern religions were developed upon the vivid transcendental states of ordinary human beings like Abraham, Moses, Christ, Mohammed, Buddha, Nanak, Joseph Smith and many more. All of those individuals around whom many religions have evolved, experienced God or some sort of Spiritual enlightenment. Each of those human beings was convinced that he had accessed the cosmic significance of the universe. Their anomalous temporal lobe functions triggered various religious movements that affected the development of human civilizations. My last book The God Parasite describes this in detail.

But the emergence of spirituality goes way back than the evolution of religions. That's why even today there are many people who consider themselves to be more spiritual than religious. Once primitive spirituality evolved beyond natural worship, it acquired roots of spirit origin but was nevertheless always conditioned by the social environment. As nature-worship developed, man's spirituality transcended into its true realm of supermortal world where there existed the nature spirits for lakes, trees, waterfalls, rain and hundreds of other ordinary terrestrial phenomena.

Feeling had controlling influence over those primitive people, not thinking. It played a crucial part in the evolutionary development of modern humans. That's why we humans are basically an emotional species alongside being the smartest one.

Our primitive ancestors looked upon the whole nature as alive, as possessed by something. They attributed life and spirit to even the inanimate objects. This is what anthropologist Sir Edward B Tylor termed as "Animism". The savages' primitive beliefs in the Spiritual Beings led to the modern sophisticated sense of spirituality that most of the modern humans including many atheists possess.

But what is "Spirituality" itself?

We must clarify this first. Spirituality is not merely an idea or concept but it is something that is resonant with the longing for permanent, eternal, everlasting peace, joy and bliss which has haunted the human beings throughout history and for which many are hungering till this day. This hunger led to the birth of all modern religions in the first place.

The notion of spirituality is wider, less concrete and less institutionally bound than religiosity. Spirituality is the heart of all religions, yet it can prevail beyond the bounds of religious rituals, that's why many non-religious people consider themselves to be spiritual.

Now we need to find out, how exactly the term "spirit" or "soul" emerged?

The primitive ancestors of humanity were unable to distinguish between dreams and waking consciousness. To their childish mind, dreams were just as real as any other waking experience. When they dreamed about deceased friends or relatives, they assumed that the dead were still alive in some spiritual form. It was all very real to the savage who would awaken from such dreams reeking with sweat, trembling, and screaming. Out of dreams, therefore, evolved the doctrine of souls and other spiritual beings, in general, a doctrine that was rational to the savages, even though it was a childish philosophy enveloped in intense and inveterate ignorance.

Figure 1.1 Sir Edward Burnett Tylor, was an English anthropologist, the founder of cultural anthropology

This doctrine grew stronger with the biological instinct of self-preservation as it effectively helped to antidote the fear, anxiety and uncertainty of death. In addition, the involuntary physical phenomenon of sneezing fueled the idea of "soul" or "spirit". Sneezing was not originally an arbitrary and meaningless custom, but the working out of a principle.

Anthropologist E. B. Tylor derived the ethnographic facts that the Zulus thought their deceased ancestors caused sneezing. Sneezing reminded the Zulus to name and praise their ancestors. They believed that the ancestors entered the bodies of their descendants when they sneezed. Ritual specialists, such as Zulu diviners, regularly sneezed as a ritual technique for invoking the spiritual power of the ancestors. Tylor concluded that these Zulu concepts and practices, were remnants of a prehistoric era in which sneezing was not merely a physiological phenomenon, but was still in the theological stage. Many of the savages also looked upon sneezing as an abortive attempt of the soul to escape from the body. Eventually, sneezing got accompanied by some common social expression, such as "bless you".

Even though, the primitive people were not at all developed in the field of intellectuality, yet they exercised their very limited intellectual powers to develop explanations of the world in which they lived.

Primitive humans thought that the soul was associated with the breath, and that its qualities could be imparted or transferred by the breath. The brave chief of a tribe would breathe upon the newborn child, thereby imparting courage. Among early Christians the ceremony of bestowing the Holy Spirit was accompanied by breathing on the candidates. The Psalmist would say: *"By the word of the Lord were the heavens made and all the host of them by the breath of his mouth."* It was long the custom of the eldest son to try to catch the last breath of his dying father.

Later on, shadow came to be feared and revered equally with the breath. The reflection of oneself in the water was also sometimes looked upon as proof of the double self, and mirrors were regarded with superstitious awe. Every now and then, many civilized persons turn the mirror to the wall in the event of death. Some backward tribes still believe that the making of pictures, drawings, models, or images removes all or a part of the soul from the body, hence such are forbidden.

Among the primitives there was even the doctrine of three or four souls. Those who held such doctrine, believed that the loss of one soul meant discomfort, two illness and three death. One soul lived in the breath, one in the head, one in the hair and one in the heart. The sick were advised to stroll about in the open air with the hope of recapturing their strayed souls. The Greeks themselves believed in three souls; the vegetative resided in the stomach, the animal in the heart and the intellectual in the head. The Eskimos believe that man has three parts: body, soul, and name.

Primitive ancestors of humanity inherited a natural environment, acquired a social environment, and imagined a spirit environment out of fear, ignorance and anxiety. During the nineteenth century, European social scientists developed different terms – fetishism, totemism, and animism for describing the original religion of humanity. Each term carried the same foundation of ignorance and fear of the primitives. Those terms are the products of the savages' inability of assessing the meaning and value of material objects. We are about to analyze them closely in the next chapter.

There has been a lot of dilemma whether to consider the theory of Animism as the origin of religion. However we can say that E. B. Tylor's Animism definitely was one of the foundational elements of Spirituality.

As time passed the primitive idea of the spirit world evolved as well, with a denser perception of something beyond the human world. Animism was one of the raw forms of Spirituality. And the urge to attain the spiritual bliss was the mother of all religions. Modern religions are the products of complex and sophisticated fantasies of individuals around the globe who were hungry for the spiritual bliss. So, their brain created the utterly blissful state of spirituality. And they made up rituals for all humanity to achieve that spiritual bliss.

* * *

CHAPTER 2
THE ROOTS OF SPIRITUALITY

During the nineteenth century, European social scientists developed different terms – animism, fetishism and totemism. These were what you can call the roots of modern spirituality. They are not directly connected to spirituality, but they prepared the soil of the human mind to develop the modern spirituality. They allowed the human beings to become spiritual beings.

But do not confuse them with the origin of religions, even though anthropologist E. B. Tylor went to huge lengths to endorse his concept of Animism as the origin of all religions. Modern religions were founded upon the experiences of several human beings with anomalous brain activity. Their craving for spirituality influenced the storyline of the hallucinatory experiences, but spirituality was not actually the cause of their mystical experiences. The cause existed in their specific brain regions, especially the temporal lobe. Animism, fetishism, and totemism were the origin of spirituality, but not of modern religions. And behind all these, were ignorance, fear and anxiety.

In terms of the primitive forms of spirituality there was really not much difference between humans and dogs. We can understand the primitive spirituality by observing the psychology of dogs.

In his popular survey of human evolution, The Origin of Civilization and the Primitive Condition of Man, John Lubbock explained that religion originated as the result of the primitive tendency to attribute animation to inanimate objects. However, we know now that religion has much more than just animism at its basis. Anyways, Lubbock observed, *"dogs appear to do the same".* As Lubbock's friend and mentor, Charles Darwin thought that religion could be explained in terms of dog behavior.

Like Lubbock, Darwin observed that dogs characteristically attributed life to inanimate objects. His dog's attention to a parasol blowing in the wind, suggested to Darwin that the animal assumed that objects were alive. In this animal psychology, therefore, nineteenth-century theorists had a basis for understanding animism as the primitive or savage propensity to attribute animation to inanimate objects.

So, technically when philosophers say that spirituality is what makes us superior to the animals, they are actually boasting the false pride of being humans. Because, if spirituality makes a species special, then even the dogs would be called as spiritual. So, basically we humans are not much different from dogs on this matter. There is nothing special or supernatural about spirituality after all.

Animism encompasses the belief that there is no separation between the spiritual and physical or material world. This belief expresses that souls or spirits exist, not only in humans, but also in inanimate objects such as plants, rocks, mountains, rivers and other entities of the natural environment. Along came the terms "fetishism" and "totemism".

The term "fetishism" was derived from the Portuguese word "feitiço", referring to nefarious instruments of magic and witchcraft. A fetish is an object believed to have supernatural powers. Initially, the Portuguese developed the concept of fetishism, referring to the objects used in religious cults by West African natives.

Contemporary Portuguese feitiço translates as charm, enchantment, juju or abracadabra, or more potentially offensive witchcraft, witchery, conjuration or bewitchment. But today, fetishism basically means "obsessive fascination". Fetishism is also characterized as a disorder when there is a pathological assignment of sexual fixation, fantasies or behaviors toward an inanimate object, frequently an item of clothing such as underclothing or a high-heeled shoe or to nongenital body parts such as the foot, armpit etc. In fact, far from being primitive, foot-fetish is very famous in the modern porn industry.

In such sexual fetishes, only through the use of that specific object can the individual obtain sexual gratification. The fetishist usually holds, rubs or smells the fetish object for sexual gratification or asks their partner to wear the object during sexual encounters.

Fetishism is a more common occurrence in males, and the causes are not clearly known yet. Fetishism falls under the general category of paraphilias, i.e. abnormal or unnatural sexual attractions.

The idea of the fetish has a particular presence in the writings of both Karl Marx and Sigmund Freud. While Marx's account of fetishism addresses the exchange-value of commodities at the level of the economic relations of production, Freud's concept of the fetish as a desired substitute for a suitable sex object explores how objects are desired and consumed.

Figure 2.1 Karl Marx

Marx presented his idea of fetishism as "commodity fetishism" in the first chapter of Capital: Critique of Political Economy (1867), at the conclusion of the analysis of the value-form of commodities, to explain that the social organization of labour is mediated through market exchange, the buying and the selling of commodities (goods and services).

Marx borrowed the concept of fetishism from The Cult of Fetish Gods (1760) by Charles de Brosses, which proposed a materialist theory of the origin of religion. Moreover, in the 1840s, the philosophic discussion of fetishism by Auguste Comte, and the psychological interpretation of religion by Ludwig Feuerbach also influenced Marx's development of commodity fetishism.

Karl Marx, Capital, Volume 1

Part I: Commodities and Money

Chapter One: Commodities, Section 4: The Fetishism of Commodities and The Secret Thereof

"A commodity is therefore a mysterious thing, simply because in it the social character of men's labour appears to them as an objective character stamped upon the product of that labour; because the relation of the producers to the sum total of their own labour is presented to them as a social relation, existing not between themselves, but between the products of their labour. This is the reason why the products of labour become commodities, social things whose qualities are at the same time perceptible and imperceptible by the senses. In the same way the light from an object is perceived by us not as the subjective excitation of our optic nerve, but as the objective form of something outside the eye itself. But, in the act of seeing, there is at all events, an actual passage of light from one thing to another, from the external object to the eye. There is a physical relation between physical things. But it is

different with commodities. There, the existence of the things quâ commodities, and the value relation between the products of labour which stamps them as commodities, have absolutely no connection with their physical properties and with the material relations arising therefrom. There it is a definite social relation between men that assumes, in their eyes, the fantastic form of a relation between things. In order, therefore, to find an analogy, we must have recourse to the mist enveloped regions of the religious world. In that world the productions of the human brain appear as independent beings endowed with life, and entering into relation both with one another and the human race. So it is in the world of commodities with the products of men's hands. This I call the Fetishism which attaches itself to the products of labour, so soon as they are produced as commodities, and which is therefore inseparable from the production of commodities."

Figure 2.2 Sigmund Freud, The Father of Psychoanalysis

On the other hand, Sigmund Freud conjectured on the symbolic origins of sexual fetishes but did not explore the meaning of fetishes systematically as he did dream contents. He treated fetishism as a deviation from the normal sexual aim of copulation leading to the release of sexual tension. In 1927, Freud published his paper on "Fetishism", in which he illustrated the analytic meaning and background of sexual fetishism.

Most of Freud's ideas of psychoanalysis emerged from his highly sexual perspective. In his paper on "Fetishism" he writes:

"In every instance, the meaning and the purpose of the fetish turned out, in analysis, to be the same. It revealed itself so naturally and seemed to me so compelling that I am prepared to expect the same solution in all cases of fetishism. When now I announce that the fetish is a substitute for the penis, I shall certainly create disappointment; so I hasten to add that it is not a substitute for any chance penis, but for a particular and quite special penis that had been extremely important in early childhood but had later been lost. That is to say, it should normally have been given up, but the fetish is precisely designed to preserve it from extinction. To put it more plainly: the fetish is a substitute for the woman's (the mother's) penis that the little boy once believed in and - for reasons familiar to us - does not want to give up."

But, these are much modern forms of fetishism, evolved in the modern times of civilization. But if we just turn the clock back to the primitive times, we'll notice that fetishism really had one heck of a hold over the primitive people. And this magical thinking along with Animism and Totemism influenced the evolution of modern Spirituality. The world of philosophy implies that Spirituality makes us different from the animals. But in fact, it's quite the opposite. We may have nurtured our perception of spirituality over time and have made it modern, but deep down, our primordial instincts of spirituality are just as same as the dogs'.

The term "totemism," according to the Scottish ethnographer John Ferguson McLennan, referred to communal alliances under the sign of an animal or an object that combined fetishism with exogamy, mixing the inability to evaluate materiality with regulations governing sexuality. In simple words, Totemism is a belief that each human has a spiritual or mystical kinship with a spirit-being, called as a "totem", such as an animal or plant. The totem is thought to interact with a given kin group or an individual and to serve as their emblem or symbol. Therefore the totem becomes a highly venerated figure of that specific kin group. This primitive idea is still alive in many cultures around the world in some form or another. For example many traditional Hindus still have their totemic belief in their "Kuldevata", that means *"deity of the kin group"*.

Totemism was really a fantastically bizarre idea. And technically it has given rise to many of the modern governing regulations of sexuality. The term "totem" is derived from the Ojibwa word "ototeman", meaning "one's brother-sister kin". The grammatical root, 'ote', signifies a blood relationship between brothers and sisters who have the same mother and who may not marry each other.

The English word totem was introduced in 1791 by a British merchant and translator who gave it a fabricated meaning. He believed that the term "totem" designated the guardian spirit of an individual or a kin group, that appeared in the form of an animal. The Ojibwa clans did indeed portray such idea by wearing animal skins. The Ojibwa named their clans after those animals that lived in the area in which they lived and appeared to be either friendly or fearful.

The first accurate report about totemism in North America was written by a Methodist missionary, Peter Jones, himself an Ojibwa, who died in 1856 and whose report was published posthumously. According to Jones, the Great Spirit had given toodaims ('totems') to the Ojibwa clans, and because of this act, it should never be forgotten that members of the group are related to one another and on this account may not marry among themselves. Thus evolved Exogamy. Hence in modern times sexual intimacy within family is a taboo. From this perspective the evolving mind of the humans created the concept of "Incest".

The long history of totemism is a hodge-podge of primitive ideas and observances. Early authors have enumerated these beliefs as :

1. Segmentation into groups conscious of their identity;
2. The bearing by each group of the name of an animal, thing, or natural phenomenon;
3. The use of this name as term of address in conversation with strangers;

4. The use of an emblem, drawn on divisional weapons and vehicles, or as personal ornament, with a corresponding taboo on the use of the emblem by other groups;

5. Respect for the patron and the design representing it;

6. A vague belief in the totem's protective role and in its value as augury.

Almost any investigator who found such a condition existing among uncivilized people would class these associated beliefs and practices as a totemic complex. But investigators needed time to understand true totemism. The main elements of true totemism were the marriage regulations and beliefs in blood relationship with the totem.

In the year 1914, one of the most famous theoretical writers on totemism, W. H. R. Rivers defined it by the coalescence of three elements:

1. A social element, viz. the connection of an animal or vegetable species, or an inanimate object, or perhaps a class of inanimate objects, with a group defined by the society, typically with an exogamous group or clan;

2. A psychological element, viz. a belief in a relation of kinship between members of the group and the animal, plant, or thing, often expressed in the idea that the human group is descended from it;

3. A ritual element, viz. respect for the animal, plant, or thing, typically manifested in a prohibition on

eating the animal or plant, or on using the object, except on certain conditions.

Even though modern religions are built upon the foundation of transcendental experiences of individuals around the world, the innate urge underneath those experiences was primitive. And basically you can find many regulatory similarities between modern religious laws and the totemic complex. Because the mind that influenced the arrival of the religions was not much different from the early human mind with totemic beliefs. Let's see how.

In the widest sense of the term, we may speak of totemism as :

1. The whole population consists of totem groups, each of those groups has a certain relationship to a class of object (totem), animate or inanimate;
2. A member of these totemic groups cannot (except under special circumstances, such as adoption) change his membership.

Here you can always conceive that the same kind of notion prevails in modern religions. In our modern society each religious group believes to have a unique relationship with its specific God, although the new generation is breaking these primitive boundaries. Also, converting to other religion is perceived as a sin in most religions. Most of the religious regulations were founded by various ordinary individuals around the world who mistook their hallucinations as the voice of God, while the primitive totemic complex was still burning bright deep down their heart.

Figure 2.3 North American native people provide one of the most recognizable examples of totemism in all of human culture—the totem pole. Totem poles are monumental sculptures carved on poles, posts, or pillars with symbols or figures made from large trees, mostly western red cedar.

However, there was one benefit of the totemic complex. It was the concept of Incest. The primordial taboo of marrying or sexual intimacy within family has proven to be beneficial from an evolutionary perspective. Marriage outside the family and even outside the totem groups brought diversity in the genetic blueprint of the progeny. And diversity has always been a very important element in the evolutionary process of all species on earth for the development of new genetic traits. Hence exogamy proved beneficial for our species.

There was way more totemic influence over the modern religious rules than we can imagine. That's because most of the religious leaders who claimed to talk to God were savages themselves. Good or bad, all the totemic complex regulations were part of the evolving mind. So those regulations have been all along the religious evolution.

For example, totemic relationship implies that members of a totem group share totemic relation among each other. And we can see the same kind of perception in Judaism. Abraham perceived that his descendants were the chosen people of the God of Israelites. It is not even a single bit advanced than the aboriginal totemic complex.

Another obligatory rule of behavior in true totemism was the prohibition on eating the totem species. And the same still goes on in the Hindu culture in the form of cow worship with its full glory of divinity. So, basically to the Hindus, eating beef is a sin, although I'm not so sure about the new generation of the Hindus. But, let's be honest, it's just another remnant of the totemic complex in the modern so-called civilized society.

In this 21st century we are living in our classy apartments in the mist of smart technological wonders. So our today's world is really what may be called a transhumanist world. Around us, we have our technological extensions. Cars are the extension of human legs; our smartphones are the extension of several parts of the human anatomy – ears, mouth and fingers (as we love texting so much instead of writing long letters with a pen). Flights are the human extension of wings. So, technically we have lost contact with nature. Modern Homo sapiens don't live in the jungle any more. So we no more have to compete with our fellow animal species.

Now look at the conditions of your primitive ancestors. They were savages who had only raw natural resources at their disposal in order to survive each day. Mother Nature was their nest. While living in the unsophisticated natural environment their natural tendency was to postulate intimate connections between man and natural beings or objects.

When McLennan launched his conception of totemism in his Fortnightly Review articles called "The Worship of Animals and Plants" we found the famous formula: *"totemism is fetishism plus exogamy and matrilineal descent"*. But it was still just a basic idea, which was yet to be understood properly. E. B. Tylor suggested *"it is necessary to consider the tendency of mankind to classify out the universe"*. He further continues,

"What I venture to protest against the manner in which totems have been placed almost at the foundation of religion. Totemism, taken up as it was as a side-issue out of the history of law, and considered with insufficient reference to the immense framework of early religion has been exaggerated out of proportion to its real theological magnitude."

So, the real matter is, Tylor introduced his idea of Animism as the foundation of religion, whereas McLennan promoted Totemism. Here we lack objectivity. But the fact is that, neither of these ideas was the foundation of modern religions. Rather, animism, totemism, fetishism all of these together were the foundation of our modern sense of spirituality. So, we may have developed our sophisticated perception of spirituality over time, but its roots remain back in the primitive days of mankind.

* * *

CHAPTER 3

THE SPIRITUAL GUIDANCE ENGINE

Humans are the most interesting and at the same time confusing creatures on planet earth. Humans can be as tough as iron and quite amazingly, they can be real weak and vulnerable in times of distress. That's why evolution has built various intricate mechanisms inside the human brain. These mechanisms are the pinnacle of biological revolution. They are the reason we can act as a "human being". They provide us the human qualities of love, passion, compassion, kindness, sympathy, empathy, rage, jealousy, envy and many more.

Among those qualities, is one that creates the inexplicable domain of divinity with all the supernatural elements - it is the Spirituality. Here we are not talking about religions at all. All the religious laws have been developed by the humans, not by any omnipresent Creator. None of the religious, spiritual and mystical experiences of human history had any divine intervention. All took place right inside the human brain, nowhere else. But the sense of spirituality has become hardwired within the cognitive structure of the human brain as it aided in the path of evolution.

297

Any cognitive structure requires a point of objectivity to work at its fullest capacity. For example, your brain has the elements to love another human being. But you cannot really feel the burning passion of love, until you find someone to burn for. And when your loving brain finds the living point of objectivity in your dearly beloved, you get flooded with the fantabulous neurotransmitters of love and pleasure – oxytocin, serotonin, dopamine and endorphins.

Likewise in some historical cases the self-sustaining cognitive structure of spirituality established the point of objectivity and named it in various cultural and social ways, like God, Jehovah, Allah, Brahma and many more. It's a big world, so there is no dearth of names for that point of spiritual objectivity. Following that point, emerged the laws and regulations from the human mind that became the foundation of modern religions.

A simple Darwinian perspective is that if there had not been any survival value associated with the experiences of spirituality and religiosity, the underlying cognitive structures should have been selected against long ago. They should have been deleted from our genetic expression. But, just like love, compassion and morality, spirituality and religiosity have made the path of survival less anxious for the humans. These evolutionary traits have acted in our favour. Naturally the hominin brain has sustained and developed the responsible cognitive structures throughout our millions of years of evolution.

Now we shall look inside the human brain and understand the various structures and microstructures responsible for developing and sustaining spirituality and religiosity. The simplest perception of modern neuroscience is that different structures inside the human brain dictate different behavioral responses in humans. All experiences, such as the sense of self, the feeling of love, the sense of spirituality, the presence of God, are generated by the functions of various neurobiological structures within the human brain. Beliefs are the consequence of these structural patterns. So, it may appear to anyone easily that spiritual/religious/mystical experiences and beliefs associated with those structures are predictable phenomena.

In modern neurotheology we examine the physical bases to spiritual, religious and mystical experiences and beliefs, that implicitly appear to reduce this rich phenomenology to only neuronal functions. So, the thing is, there was no divine intervention necessary in the evolution of spirituality and religiosity. *"Human Brain is the true God of all beliefs"*.

If structure dictates function, then the most conspicuous feature of the human brain is the two hemispheres. The total surface area is about 1570 cm^2 of which the right hemisphere contains about 40 cm^2 more area. Of the approximately 196 sulci and gyri that can be differentiated in both hemispheres only four sulci share clear structural similarity between the two hemispheres. The remainder show significant hemispheric differences. When these qualitative differences are added to the variability of the bulk volume of the cerebrum and the often forgotten fact that during ontogeny different regions mature at different rates, the complexity of structure allows for both tonic and phasic manifestations of experiences and functions.

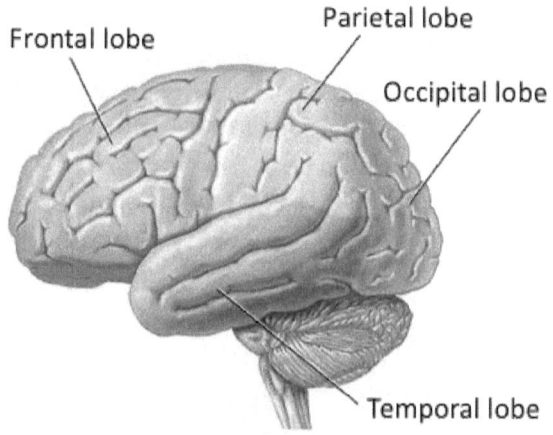

Figure 3.1 The four lobes shown in the left hemisphere of the human brain

Or in layman's terms, this complexity and variability of the brain structure make each human being different from the other. This is why every human being is always subjective about anything and everything in this universe one way or another. Every human being thinks of himself or herself to be the right one. Every human being sees things from a different perspective than the other. And the most amazing doodle of universal harmony in diversity of perspectives was seen when two great minds of human history, one scientific and the other spiritual, met in the year 1930.

Figure 3.2 When Science Met Spirituality, Einstein & Tagore

On July 14, 1930, Albert Einstein welcomed into his home in the outskirts of Berlin the Indian philosopher and Nobel laureate Rabindranath Tagore. The two proceeded to have one of the most stimulating and intellectually enthralling conversations in history, exploring the age-old friction between science and religion. The book "Science and the Indian Tradition: When Einstein Met Tagore" recounts the historic encounter, amidst a broader discussion of the intellectual renaissance that swept India in the early twentieth century, sprouting a curious fusion of Indian traditions and secular Western scientific doctrine.

The following excerpt from their conversations exudes the fascinating symposium of previously examined definitions of science, beauty, consciousness and philosophy, towards the understanding of the most fundamental questions of human existence.

"EINSTEIN: Do you believe in the Divine as isolated from the world?

TAGORE: Not isolated. The infinite personality of Man comprehends the Universe. There cannot be anything that cannot be subsumed by the human personality, and this proves that the Truth of the Universe is human Truth. I have taken a scientific fact to explain this — Matter is composed of protons and electrons, with gaps between them; but matter may seem to be solid. Similarly humanity is composed of individuals, yet they have their interconnection of human relationship, which gives living unity to man's world. The entire universe is linked up with us in a similar manner, it is a human universe. I have pursued this thought through art, literature and the religious consciousness of man.

EINSTEIN: There are two different conceptions about the nature of the universe: (1) The world as a unity dependent on humanity. (2) The world as a reality independent of the human factor.

TAGORE: When our universe is in harmony with Man, the eternal, we know it as Truth, we feel it as beauty.

EINSTEIN: This is the purely human conception of the universe.

TAGORE: There can be no other conception. This world is a human world — the scientific view of it is also that of the scientific man. There is some standard of reason and enjoyment which gives it Truth, the standard of the Eternal Man whose experiences are through our experiences.

EINSTEIN: This is a realization of the human entity.

TAGORE: Yes, one eternal entity. We have to realize it through our emotions and activities. We realized the Supreme Man who has no individual limitations through our limitations. Science is concerned with that which is not confined to individuals; it is the impersonal human world of Truths. Religion realizes these Truths and links them up with our deeper needs; our individual consciousness of Truth gains universal significance. Religion applies values to Truth, and we know this Truth as good through our own harmony with it.

EINSTEIN: Truth, then, or Beauty is not independent of Man?

TAGORE: No.

EINSTEIN: If there would be no human beings any more, the Apollo of Belvedere would no longer be beautiful.

TAGORE: No.

EINSTEIN: I agree with regard to this conception of Beauty, but not with regard to Truth.

TAGORE: Why not? Truth is realized through man.

EINSTEIN: I cannot prove that my conception is right, but that is my religion.

TAGORE: Beauty is in the ideal of perfect harmony which is in the Universal Being; Truth the perfect comprehension of the Universal Mind. We individuals approach it through our own

303

mistakes and blunders, through our accumulated experiences, through our illumined consciousness — how, otherwise, can we know Truth?

EINSTEIN: I cannot prove scientifically that Truth must be conceived as a Truth that is valid independent of humanity; but I believe it firmly. I believe, for instance, that the Pythagorean theorem in geometry states something that is approximately true, independent of the existence of man. Anyway, if there is a reality independent of man, there is also a Truth relative to this reality; and in the same way the negation of the first engenders a negation of the existence of the latter.

TAGORE: Truth, which is one with the Universal Being, must essentially be human, otherwise whatever we individuals realize as true can never be called truth – at least the Truth which is described as scientific and which only can be reached through the process of logic, in other words, by an organ of thoughts which is human. According to Indian Philosophy there is Brahman, the absolute Truth, which cannot be conceived by the isolation of the individual mind or described by words but can only be realized by completely merging the individual in its infinity. But such a Truth cannot belong to Science. The nature of Truth which we are discussing is an appearance – that is to say, what appears to be true to the human mind and therefore is human, and may be called maya or illusion.

EINSTEIN: So according to your conception, which may be the Indian conception, it is not the illusion of the individual, but of humanity as a whole.

TAGORE: The species also belongs to a unity, to humanity. Therefore the entire human mind realizes Truth; the Indian or the European mind meet in a common realization.

304

EINSTEIN: The word species is used in German for all human beings, as a matter of fact, even the apes and the frogs would belong to it.

TAGORE: In science we go through the discipline of eliminating the personal limitations of our individual minds and thus reach that comprehension of Truth which is in the mind of the Universal Man.

EINSTEIN: The problem begins whether Truth is independent of our consciousness.

TAGORE: What we call truth lies in the rational harmony between the subjective and objective aspects of reality, both of which belong to the super-personal man.

EINSTEIN: Even in our everyday life we feel compelled to ascribe a reality independent of man to the objects we use. We do this to connect the experiences of our senses in a reasonable way. For instance, if nobody is in this house, yet that table remains where it is.

TAGORE: Yes, it remains outside the individual mind, but not the universal mind. The table which I perceive is perceptible by the same kind of consciousness which I possess.

EINSTEIN: If nobody would be in the house the table would exist all the same — but this is already illegitimate from your point of view — because we cannot explain what it means that the table is there, independently of us. Our natural point of view in regard to the existence of truth apart from humanity cannot be explained or proved, but it is a belief which nobody can lack — no primitive beings even. We attribute to Truth a super-human objectivity; it is indispensable for us, this reality which is

independent of our existence and our experience and our mind —
though we cannot say what it means.

TAGORE: Science has proved that the table as a solid object is
an appearance and therefore that which the human mind perceives
as a table would not exist if that mind were naught. At the same
time it must be admitted that the fact, that the ultimate physical
reality is nothing but a multitude of separate revolving centres of
electric force, also belongs to the human mind. In the apprehension
of Truth there is an eternal conflict between the universal human
mind and the same mind confined in the individual. The perpetual
process of reconciliation is being carried on in our science,
philosophy, in our ethics. In any case, if there be any Truth
absolutely unrelated to humanity then for us it is absolutely non-
existing. It is not difficult to imagine a mind to which the sequence
of things happens not in space but only in time like the sequence of
notes in music. For such a mind such conception of reality is akin
to the musical reality in which Pythagorean geometry can have no
meaning. There is the reality of paper, infinitely different from the
reality of literature. For the kind of mind possessed by the moth
which eats that paper literature is absolutely non-existent, yet for
Man's mind literature has a greater value of Truth than the paper
itself. In a similar manner if there be some Truth which has no
sensuous or rational relation to the human mind, it will ever
remain as nothing so long as we remain human beings.

EINSTEIN: Then I am more religious than you are!

TAGORE: My religion is in the reconciliation of the Super-
personal Man, the universal human spirit, in my own individual
being."

Now, that you have surfed the ocean of two genius
minds, let's explore further the physical basis of the

spiritual oasis in terms of hemispheric structures of the brain.

In general the left hemisphere in most individuals is associated with sequential integration of experiences and consequently is strongly enmeshed with language processes. The sense of self is strongly correlated with language and consequently with the left hemisphere. The right hemisphere which is structurally organized for more simultaneous integration of experiences displays some linguistic properties, whose syntactic and semantic levels are estimated to be at the level of a young child and adolescent, respectively. In the right hemisphere, information is organized as a function of vigilance, anticipation and emotion, with the purpose of seeing "the big picture".

Or in simple words, the left hemisphere of your brain makes you aware of yourself, while the right hemisphere is always anticipating something out of the ordinary in its surrounding environment. That's why even a slight rushing of wind against your skin in a dark scary environment can give you goose bumps and make you sense as if something paranormal has just passed by you. This kind of sense of a paranormal presence is what we call "sensed presence". And that's why if you are a person of weak heart, you should avoid visiting a graveyard at night, because the right hemisphere of your brain would already be anticipating ghosts and any subtle change in the environment can scare the hell out of you. The evolutionary purpose of such brain function is to foresee danger and avoid it.

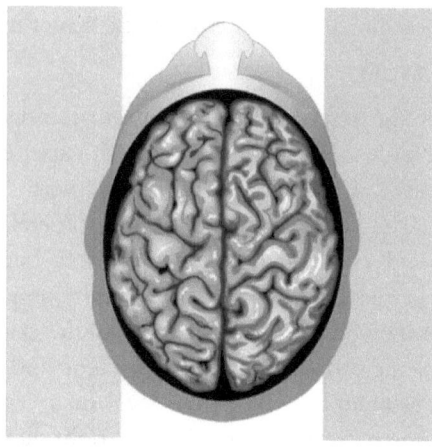

Figure 3.3 The Left Hemisphere is responsible for the sense of self and The Right Hemisphere is responsible for the sense of the supernatural

The caudal region of the cerebral hemispheres is the general reservoir for the perception and understanding of experiences which are the intermediate representations of neuronal activity generated by sensory patterns evoked by physical stimuli. They are a small subset of the billions of events that penetrate the cerebrum every second. The frontal region organizes and re-represents these caudal experiences and results in the person acting upon the environment. Beliefs, attitudes, fantasies that will occur within the future, world views about one's self and one's relationship to others, all are strongly correlated with the integrity of this region.

The right hemispheric equivalent to the left hemispheric sense of self is more related to the feeling of "another" that can be evoked experimentally. So, the bottom-line is that we can experimentally evoke experiences of God, ghosts, aliens and angels in a person who believes in them. When there are intermittent intrusions of the electromagnetic patterns (action potentials of neurons) into left hemispheric processes from the right hemispheric equivalent of the self, the person experiences a sensed presence, the feeling of a Sentient Being. Experiments have shown that the burst spiking of a single cortical neuron is sufficient to modify the entire brain state. The activity of a single neuron can initiate a cascade that changes the probability of the occurrence or non-occurrence of a response. So any slight sensory stimulus from the slightest change in the environment can evoke a flood of neural activity within the brain and turn on an experiential response of spirituality, religiosity and mysticism. In that state of mind, the experience seems to be the most real thing in the entire universe.

Then comes the labeling part of the person's experience. The labels in this case are acquired through culture and are applied to these transient experiences. These labels actually determine the details of the verbal images and their significance, when the occurrences are later reconstructed through the autobiographical processes involved primarily with the right prefrontal regions and temporal lobe structures (hippocampus and amygdala). The languages of all cultures have default descriptors for "everything and everywhere", that are the essential operations for gods or divinity, with which the experiences are associated. In this manner the sense of the person becomes adamantly interlaced with the shared beliefs of the culture.

Whether a person shows affinity for a specific God of a specific religion or just considers himself or herself to be simply spiritual but not religious, the brain function underneath both the system of beliefs is mostly the same. It's the brain's own guidance engine, a coping mechanism that promotes self-preservation.

The fundamental characteristics of Spirituality or belief in a Supreme Spiritual Being reflect their origin in the right hemisphere. The basic sensation is that it is something beyond. It feels something familiar yet something out of this world and more than the "self". This may sometimes produce the experience that *"the self is complete now"*. The themes are dominated by emotional significance, a perfusion of the sense of self with simultaneity. Now you might ask, what the heck is "simultaneity"? Simultaneity is the ultimate state of the mind where a person feels being outside space and time. Often this experience of simultaneity includes being one with the universe and the sensation of utter meaningfulness of the universe.

From a neuroanatomical perspective, the dominating theme of the experience would reflect which region of the right hemisphere dominates input through the major interhemispheric pathways: the corpus callosum, anterior commissure, or dorsal hippocampal commissure. Parietal (spatial and location experiences), anterior temporal (auditory, vestibular, "knowing" experiences), posterior temporal (visual "manifestations"), insular (visceral, including sexual experiences) and hippocampal-amygdala ("remembering" and "intrusive memories") sources differentially weigh the details of the experience.

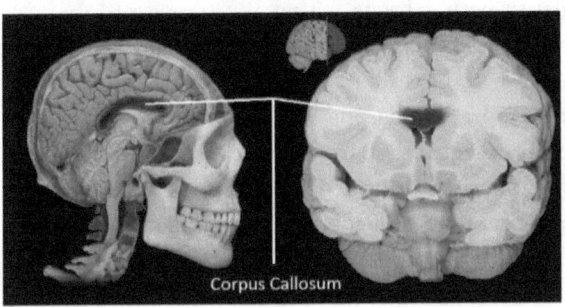

Figure 3.4 The corpus callosum, present only in mammals, is the largest fiber tract in the primate brain. This interhemispheric fiber system is composed of approximately 200 million axons.

There is an utter importance of the right hemispheric sense of the supernatural. Let's figure it out. The occurrences of spiritual and religious experiences are more probable when the right hemispheric intrusions are more frequent, such as while dreaming or in the fluctuating waking states during the early morning hours. During more stimulant conditions, such as psychological depression, despondency and hopelessness, the intrusion of right hemispheric processes and subsequent activation of the left hemisphere are associated with both a sensed presence and the sense of powerful personal significance. And this leads to the recovery from the depression. Now, let's have a look at the extreme state of the religious and spiritual experiences.

311

The extreme spiritual and religious experiences that the religious leaders had back in the days, were also the product of interhemispheric intercalation. Any electrically labile condition, such as partial complex epilepsy with a focus within the temporal lobes, would facilitate interhemispheric intercalation. The behaviors of many if not all of the individuals who began the contemporary religions meet the criteria for the ictal (during electrical paroxysmal activity) and interictal behaviors of temporal lobe epilepsy. This is what we call "temporal lobe personality". And this kind of conscious awareness of the supernatural world is what we call "parasitic consciousness", from which the title of my last book emerged, "The God Parasite".

The emergence of the "temporal lobe personality" defined by enhanced religiosity, spirituality, paranormal interests, hypermoralism, altered sexuality, a sense of the personal meaningfulness, a feeling of being "selected" and the compelling motivation to evangelize are classic clinical correlates. In many cases, psychotic behaviors emerge about two decades after the onset of the electrical seizures within the temporal lobes.

Such electrical activity within the temporal lobe can suddenly turn an atheist into a believer. We have many clinically recorded cases of such strange conversions. One file mentions of a middle aged man who did not believe in a god. During various stages of the subsequent disorder he thought he was Jesus who had returned to earth to preach and to spread the word by starting a musical group. Later he believed God was speaking through the radio and experienced a *"revelation"*

312

that heaven was a secret place hidden in Siberia. Following a partial lobectomy of the right temporal lobe no overt convulsions occurred and his receptive language skills normalized. Although he still pondered the existence of a god, he no longer confused the differentiation between his thoughts and the concept.

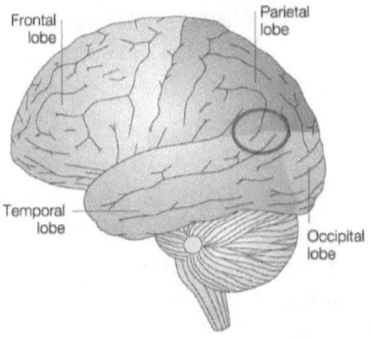

Figure 3.5 Left Temporoparietal Junction showed in circle

The capacity to reason about the supernatural and spiritual beliefs requires the normal integrity of the left temporoparietal junction. We'll elucidate further on this capacity in the next chapter.

Now let's have a real life approach to the spiritual and religious development phenomenon of the human mind. Many atheists, without any proper scientific citation suggest that teaching religion in school is not beneficial for the growing mind of a kid. But the fact is, our brain is designed in such a way that even if the religious laws are not taught to a kid, in times of distress and depression, he or she would develop his or her own beliefs in some kind of Spirituality or a Supernatural

Entity. To understand this phenomenon more clearly let's carry out a thought experiment. Shall we! In this thought experiment let's create a hypothetical situation where a boy and a girl grow up together without any parental guidance and technological assistance.

Two children (brother and sister, let's call them Adam and Eve) survive a shipwreck and get marooned in a lush tropical island. In the island they have no social barriers and regulations of any kind whatsoever. They are all on their own, with no prior religious teachings. All they have to guide themselves on this island is their biological instinct. And the biological instinct of any living species on planet earth is self-preservation. Years pass and both of them become tall, attractive, and beautiful young adults. They live in their primitive hut, spending their days together fishing, swimming and diving for pearls. Being completely ignorant about the regulations of sexual intimacy, their relationship of brother and sister blossoms into romantic love. Their flourished biology attracts each other sexually and eventually they engage in passionate sexual intercourse for the first time. Their brains find the pleasure in love making, so now they regularly make love.

One day, Eve steps on a stonefish and becomes seriously ill. Adam gets really worried. Finding no way to cure his partner, Adam's biological instinct drives his mind to recognize a giant rock of absurd shape as some sort of Supernatural Entity that has abilities beyond humans. So he starts worshipping it and eventually Eve recovers on her own. Naturally, the primitive mind of Adam would give credit for Eve's recovery to the mystical nature of that giant stone. Now they start worshipping it as a daily ritual of good omen.

Eventually Eve becomes pregnant and one night she gives birth to a baby boy. Aggravated at not knowing how to feed the baby, she holds him close and the baby instinctively starts suckling on her breast. The young parents spend time playing with their child, as he grows. They teach him how to swim, fish and build things. And life goes on like this.

An Illustrative Flowchart of Biological Instincts:

1.

2.

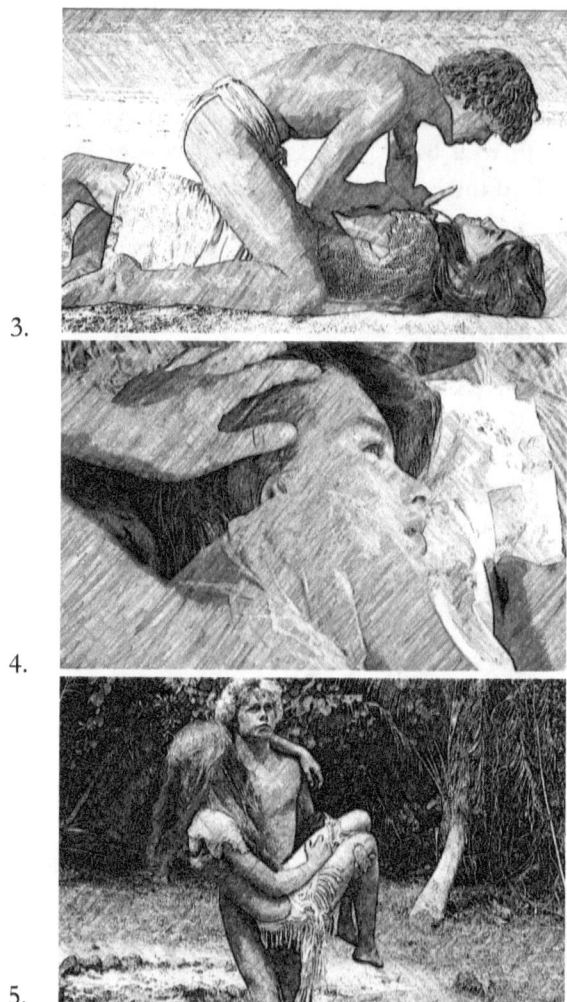

3.

4.

5.

6.

7.

8.

9.

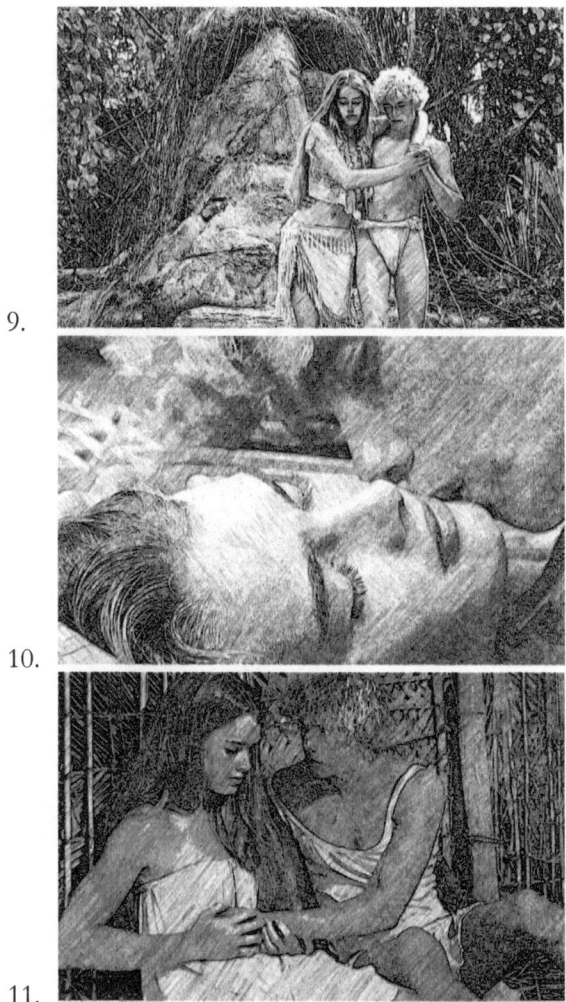

10.

11.

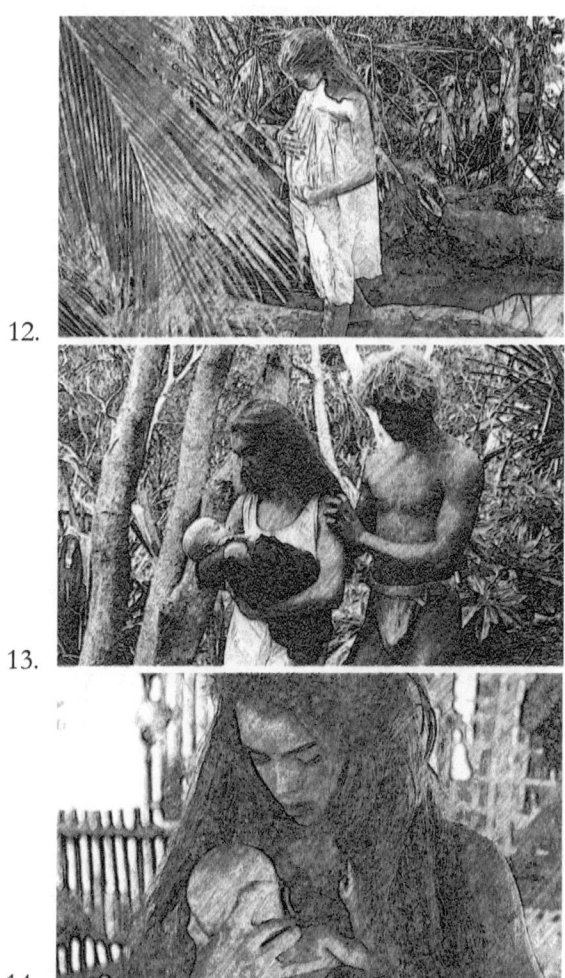

12.

13.

14.

15.

16.

In a battle between "Nature and Nurture", it's always "Nature" that wins. It's the basic rule of biological life on earth. No matter how much you nurture a cub and try to make it a pet by feeding it milk and food, its biological instinct will force it to hunt its prey. And when it grows big, it'll quench its hunger by hunting anything in the grasp of its paws. A tiger will remain a tiger, no matter how much effort you make in taming it. Likewise, a Human will remain a Human, with all the emotions and beliefs. Human emotions and beliefs act as an in-built GPS that guides them towards the path that's best for self-preservation.

*   *   *

CHAPTER 4
THE SPIRITUAL IMMUNE SYSTEM

Let's talk about love!!!

The most beautiful concept on our planet!!!

When you fall for someone, you become absolutely blind to the imperfections and negative characteristics of that special person. It feels as if that one person is the most perfect and righteous being in this entire universe. And if someone raises a finger at that person's some bad qualities, your rage would reach the top of the Mount Everest, because you are immune to the imperfections of your beloved one. That's your brain's immune system to protect your relationship, because being in love with someone gives you an inexplicable bliss. Anything that activates the pleasure centers of your brain, takes away a lit bit of your reasoning abilities.

To be honest, when you are in love with someone, who cares about logic, even though you know very well that the sensation of love is the product of a beautiful interplay of neurotransmitters in the brain. All you can think of is the euphoric sensation you experience when that someone is next to you. Everything seems so pleasurable.

The same goes for Spirituality. Human brain is structured to avoid any kind of refutation of one's spiritual and religious beliefs. The functional utility of belief in Spirituality and some sort of God is persistent in human history. Such consistency throughout time strongly suggests that spiritual and religious beliefs reflect a property of the human brain and are associated with particular configurations of cerebral activity and cerebral structures.

One of the most important secular explanations for the persistence of Spirituality and Religiosity is that they reduce the potentially incapacitating anxiety associated with our ability to anticipate the dissolution of the self, i.e. personal death. Any thought sequence that minimizes anxiety would get reinforced over time. The sequence would be maintained during the person's development and would be provoked in contexts where refutation of the belief might occur. So, any kind of refutation of a person's spiritual or religious beliefs makes the beliefs only stronger. Repudiation reinforces beliefs. Which means, atheistic approach towards a spiritual or religious person would only strengthen his or her spirituality or religiosity.

Well, now the road is going to be a little bumpy with neuroscientific technicality. So, fasten your sit belt.

The capacity to reason about the beliefs of others requires the normal integrity of the left temporoparietal junction. This region has been also implicated in the theory of mind. Theory of Mind refers to the cognitive capacity to understand mental states such as beliefs, intents, desires, pretending, knowledge etc. of oneself and others The medial prefrontal cortices, immediately adjacent to the region involved with moral decision making, is also consistently activated by "Theory of Mind"-related tasks in which people think about their own or others' mental states. The labels for what these states mean or imply are strongly influenced by cultural conditioning of the person associated with those words.

As for proper social functioning, any kind of damage or dysfunction of the orbitomedial and polar frontal cortices is often associated with acquired sociopathy. The dynamic functions of the ventromedial frontal lobe that involve in social functioning have been only recently accessible by modern imaging techniques such as fMRI (functional Magnetic Resonance Imaging). This region is necessary for the normal generation of social emotions, in conjunction with moral judgments of right and wrong. Lesion studies, have shown that damage in this region results in deficits of social conduct and strikingly reduced compassion, shame, and guilt, all of which are closely associated with moral values.

The ventromedial frontal lobe involves the gyrus rectus, orbitofrontal cortices, anterior cingulate cortices and ventral cingulate cortex. Damage to the ventromedial prefrontal cortices (particularly in the right hemisphere) results in an abnormally "utilitarian" pattern of judgments on moral dilemmas that juxtaposes considerations for the welfare of the aggregate against aversive behaviors. The most typical example of the later would be killing a smaller subset of people to save a larger number of others. Because of the involvement of the anterior cingulate which is involved with bonding and love, particularly to the group, this affiliation can influence the bifurcation sequences that determine the outcome of the logical argument. The enhancement of this bonding during late adolescence to early adulthood intertwines the identity of the self with the emotionally bonded group and the verbal behavior. The later is derived from the essential beliefs that define the group.

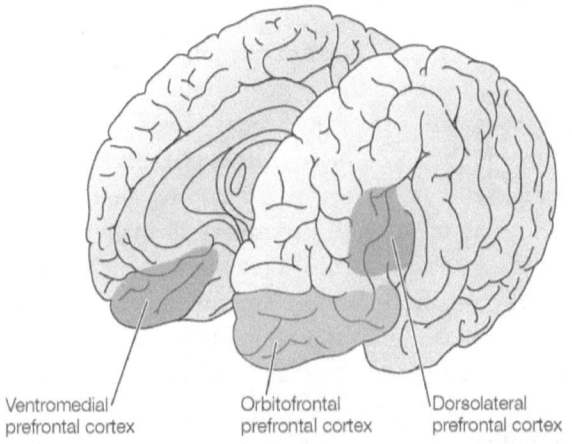

Figure 4.1 Rational decision making zones of the brain

The participation of the orbitofrontal region which associates the history of rewards and punishments with the consequences of planned thoughts and actions would create conditions for moral justification based upon beliefs in general and religious beliefs in particular. The major input from the amygdala, which is associated with the degree of manifestation of overt (killing others) or covert (self-destructive acts) aggression, into the orbitofrontal regions saturates the behavior with meaning and potentially moral justification. Because behavior is reinforced by its consequences, aggression exhibits a low threshold for many individuals for which the belief is maintained and through which frustrations induced by challenge of the belief is expressed. So, any kind of sense of threat to an utterly spiritual or religious person's belief would only make that person furious, no matter how much reasonable argument is made against the belief.

The sense of self is a primary linguistic process that evolves as reiterative integrations of the verbal representations of experiences over the person's lifetime. If the sense of self and its beliefs of an infinite spiritual state are indoctrinated in early childhood, then by the time the person progresses through the various cognitive stages and exhibits the potential perceptiveness to refute the argument, even its initiation would be anxiogenic (something that causes anxiety). In one experimental study by Dr. Michael Persinger of the Laurentian University normal volunteers who were exposed to progressive statements that would ultimately suggest God was a creation of neuronal function, displayed behavioral indications that the right hemisphere discerned the semantic implicit chain even without the participation of awareness. The cognitive sequence that would have led to the conclusion was avoided or displaced. Ergo, human brain is hardwired to avoid any kind of refutation of spiritual and religious beliefs.

The right temporal lobe in general and the hippocampus in particular, in addition to displaying the lowest threshold for electrical excitability and seizure-production, is remarkably sensitive to ambient geomagnetic activity. This occurs particularly during dreaming, which is characterized by an enhancement of activity within the right temporal region. This counts for the so-called spiritual or religious experiences in dreams. Modification of experiences mediated through the right hippocampus into the left equivalent can modify memory even without the person's awareness. The brain takes care of everything that is somehow beneficial to the human being, one way or another.

The left hemispheric functions are associated with the sense of self and the experience of conscious awareness. And the right hemisphere is configured to monitor subtle changes within the environment, particularly visuospatial patterns, meaning and emotional significance. We often call this phenomenon "the hidden observer" effect.

Now here is a very interesting biological impact of spiritual belief upon human life. In an experiment it has been found that people who tend to possess a spiritual personality and like to live life in a spiritual manner, display blood chemistry typical of right hemispheric dominance. Spiritually inclined individuals relative to atheistic ones, exhibit increased serum levels of serotonin, quinolinic acid and nicotine, but decreased level of dopamine, noradrenalin and morphine. Increased level of serotonin encourages normal mood, while nicotinic cholinergic stimulation can facilitate vigilance. Diminished levels of dopamine and morphine suggest that spiritual beliefs, particularly rituals can replace the requirement for addictive substances.

So, basically experimental data imply there are various intricate neuromechanisms in the brain by which an average religious or spiritual person avoids or ignores the logical conclusion that his or her belief in the supernatural may not be valid. Every time the religious or spiritual belief is provoked, to reduce anxiety the strength of the belief is more strengthened by the neurobiology itself. Effectively in human life the belief becomes a powerful cognitive anxiolytic and antidepressant for the daily challenges.

* * *

CHAPTER 5
EXTREMISM & ATHEISM

Alongside the positive aspects there are negative elements to this intellectual opiate of the masses. The human species is one of the most aggressive species on earth that has ever existed. We are the Tyrannosaurus Rex of the mammals. As a species we have killed every single animal form that we have encountered. Every single discovery we have made from gunpowder to atomic fission has been to kill a perceived enemy. Since the year 1500, approximately 150 million human beings like you and me, have died during armed conflicts between nations.

Throughout history various cultures around the world have raped and slaughtered in the names of Jehovah, Allah, Rama or the Great Cosmic Guide. Even though sociologists and historians often try to attribute these episodes to political or economic causes, the individuals who engaged in killing often reported that they felt "God's consent" in their great activity. They were mentally unstable individuals who used religion as the justification for their monstrosity. But don't even for a second think that every person who possesses spiritual or religious beliefs might kill others if they believed their

God had commanded to do so. In an experiment only 7% of the volunteers replied "Yes" to the statement *"If God told me to kill I would in His Name"*.

Most religious texts, including the Bible, Quran, the Book of Mormon, and many others have frequent references to violent killings as the proof of God's power of justice. Suppose another group of people exists who embraces a different God. If another God exists then does your God loose its omnipotent power? What happens when the experiences attributed to this God indicate that he or she is the only one God and believers in any other deity are less than human and hence expendable? So, as a basic religious response people who do not belong to the same religious or spiritual belief system are marginalized as unenlightened and become candidates for extinction. Nomad groups believing themselves to be the chosen ones have marched into lands belonging to others and have slaughtered entire race because of their belief: *"we are the chosen people of God and he gave us the land"*. They sustained their imperialist behavior through the Ten Commandments. The omnipotence of their God is seen in the second of the Ten Commandments: *"Thou shalt have no other gods before me"*.

The same kind of self-righteousness goes on in the Quran.

Sura 112 : Al-Ikhlas

> *Qul huwa Allahu ahad*
>
> *Allahu assamad*

Lam yalid walam yoolad

Walam yakun lahu kufuwan ahad

Translation:

Say, "He is Allah, [who is] One,

Allah, the Eternal Refuge.

He neither begets nor is born,

Nor is there to Him any equivalent."

Again in the Hindu scripture Chandogya Upanishad 6:2:1 we find the phrase "Ekam evadvitiyam", that means "He is One only without a second".

This kind of self-righteous phrases are spread around the world in various religious texts that keep the road of interpretation open for people. And every single human being is different in his or her own way. That means the same phrase is interpreted in a variety of ways by different people. And some of those people take the words of thousands of years old texts so literally that they might just perceive that the God mentioned in their religious scripture is the only Supreme Lord of the Universe. This gives them the idea that all other Gods mentioned in other texts are imposters. And this leads to conflict of interest.

To quote from my last book The God Parasite: Revelation of Neuroscience

"The temporal lobes are highly interconnected with the limbic system, especially the amygdala. Amygdala is basically our emotion and fear center. When the temporal lobe acts anomalously, it triggers excitement in the amygdala, which often leads to a feeling of unearthly bliss and absolute meaningfulness of the universe. Such feeling often overwhelms an individual during the God encounter. Everything around him/her is imbued with cosmic significance. Individuals often exclaim with utmost joy, *"I finally understand what it's all about. This is the moment I've been waiting for all my life. Suddenly it all makes sense."* or, *"finally I have insight into the true nature of the cosmos"*...

Specific electrical anomalies in the amygdala during the temporal lobe storms can also trigger aggression and homicidal urges. This counts for the outrageous and murderous commands from man's God."

Therefore "Spirituality" and "Religiosity" have both positive and negative components. It's pretty simple. Say, when you are in love you can hardly see things reasonably. Likewise religious extremists can hardly perceive things with their logical mind whenever their beliefs are questioned. And if we talk in terms of recent times, then to be straight-forward, I must say that most peace-loving people from the Islamic culture are forced to pay for the actions of a few Islamic extremist communities.

There are various other harmful elements of the man-made religious regulations. The scriptures are flooded with clear depictions of megalomania, imperialism, superciliousness, short-temper, jealousy, sexism, polygamy, bigotry, savagery and prejudice. Most of the early scriptures like the Hebrew Bible, Manusmriti (also known as Manu Samhita), Vedas, Quran etc. are the specimens of human literature that are the fusion of philosophical goodness, utter stupidity and intellectual bulimia.

Now let's look into some of the real negative elements created by man in the form of religious regulations.

The Old Testament, Exodus 31:14

"Observe the Sabbath, because it is holy to you. Anyone who desecrates it is to be put to death; those who do any work on that day must be cut off from their people."

So the simple meaning is that if you see that your neighbor is working on Sunday, you should kill him.

The Old Testament, Deuteronomy 22:22-24

"If a man is found lying with a married woman, then both of them shall die, the man who lay with the woman, and the woman; thus you shall purge the evil from Israel. If there is a girl who is a virgin engaged to a man, and another man finds her in the city and lies with her, then you shall bring them both out to the gate of that city and you shall stone them to death; the girl, because she did not cry out in the city, and the man, because he has violated his neighbor's wife. Thus you shall purge the evil from among you."

Quran, Sura 4 An-Nisa

4:3

"Wa-in khiftum alla tuqsitoo feealyatama fankihoo ma tabalakum mina annisa-i mathna wathulathawarubaAAa fa-in

khiftum alla taAAdiloo fawahidatanaw ma malakat aymanukum thalika adnaalla taAAooloo"

Translations

And if you fear that you will not deal justly with the orphan girls, then marry those that please you of [other] women, two or three or four. But if you fear that you will not be just, then [marry only] one or those your right hand possesses. That is more suitable that you may not incline [to injustice].

4:34

"Arrijalu qawwamoonaAAala annisa-i bima faddalaAllahu baAAdahum AAala baAAdin wabimaanfaqoo min amwalihim fassalihatu qanitatunhafithatun lilghaybi bima hafithaAllahu wallatee takhafoonanushoozahunna faAAithoohunna wahjuroohunnafee almadajiAAi wadriboohunna fa-in ataAAnakumfala tabghoo AAalayhinna sabeelan inna Allaha kanaAAaliyyan kabeera"

Translations

Men are in charge of women by [right of] what Allah has given one over the other and what they spend [for maintenance] from their wealth. So righteous women are devoutly obedient, guarding in [the husband's] absence what Allah would have them guard. But those [wives] from whom you fear arrogance - [first] advise them; [then if they persist], forsake them in bed; and [finally], strike them. But if they obey you [once more], seek no means against them. Indeed, Allah is ever Exalted and Grand.

Or in simple terms, according to the laws given by Muhammad in a transcendental state, women must be submissive to men. As if somehow, men are superior to women. These laws were just the representation of his own manly sexual desires with no God element involved.

Now let's look behind the curtains of Hinduism, that is often considered to be relatively moderate than other religions. But the fact is that, it is no different from the others. In many cases, Hinduism is actually way worse than other religions. In all Hindu scriptures women are treated as commodities. As per Hindu scriptures after the death of her husband the wife either has to lead a life of celibacy or has to burn alive on her husband's pyre. This is what is called "Sati pratha". While on the other hand the husband is free to marry another wife after the death of his wife and can marry many wives even when the wife is alive.

Rig Veda 10:XVIII:7-8

"Let these unwidowed dames with noble husbands adorn themselves with fragrant balm and unguent. Decked with fair jewels, tearless, free from sorrow, first let the dames go up to where he lieth.

Rise, come unto the world of life, O woman: come, he is lifeless by whose side thou liest. Wifehood with this thy husband was thy portion, who took thy hand and wooed thee as a lover."

Vishnusmriti XXV:14

"After the death of her husband, to preserve her chastity, or to ascend the pile after him."

Let's see some of the verses from the Hindu text Manusmriti or Manu Samhita.

Manusmriti: The Laws of Manu

IX:72

"Though (a man) may have accepted a damsel in due form, he may abandon (her if she be) blemished, diseased, or deflowered, and (if she have been) given with fraud."

VIII:1

"If a wife, proud of the greatness of her relatives or (her own) excellence, violates the duty which she owes to her lord, the king shall cause her to be devoured by dogs in a place frequented by many."

People know of the Vedas as holy, but in the name of preaching Vedas the Brahmins tyrannized the whole Hindu culture by the holocaust of female infanticide. Female infanticide was sanctioned right from the early times, by the Vedas. Rig Veda itself states that a woman should beget sons. The newly married wife would be blessed that she could have ten sons. So much so, that for begetting a son, Vedas prescribed a special ritual called 'Punsawan sanskar'. It is a ceremony performed during the third month of pregnancy. In that ceremony it is prayed:

Atharva Veda 6.XI

"Asvattha on the Sami-tree. There a male birth is certified. There is the finding of a son: this bring we to the women-folk.

The father sows the genial seed, the woman tends and fosters it. This is the finding of a son: thus hath Prajāpati declared.

Prajāpati, Anumati, Sinivāli have ordered it. Elsewhere may he effect the birth of maids, but here prepare a boy."

Female infanticide helped the Brahmins to maintain control over conquered populations by reducing the number of women. But it didn't stop there. The Brahmins carried on their ritual of killing baby girls. They would cut the baby into pieces and then feed it to animals. Or they would throw the baby into the Ganges River where crocodiles would quickly gobble it up. The Rajputs would throw their baby girl up in the air and slice her with swords as she fell down to the ground. And to speak about modern times, it even gets amazingly fascinating in terms of ruthlessness. Today not only do we have female infanticide, we also have female foeticide (killing it in the womb). Since Indian independence, more than 50 million baby girls have been killed in India. It is an average of 1 million babies per year, and it is continuing.

Till this date, there is a custom whereby the mother asks her son on the eve of his marriage, *"Where are you going?"* He replies, *"I'm going to bring you a maid-servant".* It comes from those days when the victorious clan would drag the defeated women back to its own hill. The women would be brought back in those days with chains on their wrists, otherwise they might run away. The so called sacred bangle that a married Hindu woman wears today is a symbol of that early servitude.

Back in those days, Aryan women were covered from head to toe, and the indigenous Sudra (lower-caste) women were forced by the Brahmins to go around topless. Manu himself characterized women as ornaments.

Manusmriti IX:17

"(When creating them) Manu allotted to women (a love of their) bed, (of their) seat and (of) ornament, impure desires, wrath, dishonesty, malice, and bad conduct."

But it doesn't end there. All the so-called "sacred" Hindu texts are male-dominated, because they didn't come from any Supreme Omnipotent Entity at all. They were all laws created by simple control-freak men.

Manusmriti IX:1-3

"I will now propound the eternal laws for a husband and his wife who keep to the path of duty, whether they be united or separated.

Day and night woman must be kept in dependence by the males (of) their (families), and, if they attach themselves to sensual enjoyments, they must be kept under one's control.

Her father protects (her) in childhood, her husband protects (her) in youth, and her sons protect (her) in old age; a woman is never fit for independence."

In another Hindu text Atharva Veda one verse states that *"a wife is given by God to a husband to serve him and to bear him children. Further she is referred to by her husband as his subordinate and slave"* (Atharva Veda 14.01.52). To be honest, Hinduism is the only religion where women are considered as born promiscuous.

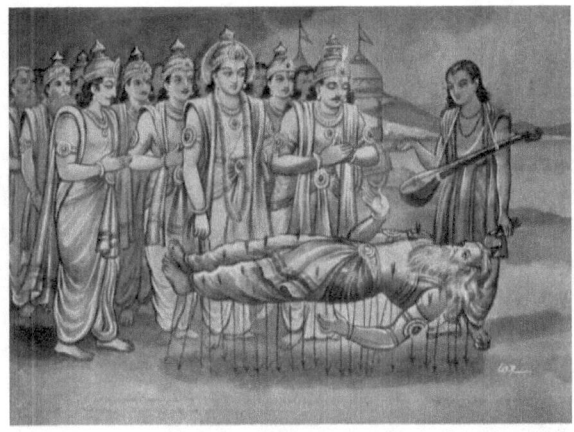

Figure 5.1 Bhishma on the bed of arrows preaching the Pandavas along with Lord Krishna and the celestial Sage Narada

In the celebrated Hindu epic Mahabharata there is a whole section where Bhishma explains to Yudhisthira the innate qualities of women by reciting a conversation between the celestial Sage Narada and the Apsara Panchachuda.

Mahabharata, Anusasana Parava, Section XXXVIII

"Yudhishthira said, 'O best of the Bharatas, I wish to hear thee discourse on the disposition of women. Women are said to be the root of all evil. They are all regarded as exceedingly frail.'

Bhishma said, 'In this connection is cited the old history of the discourse between the celestial Rishi Narada and the (celestial) courtezan Panchachuda. Once in ancient times, the celestial Rishi Narada, having roamed over all the world, met the Apsara Panchachuda of faultless beauty, having her abode in the region of Brahman. Beholding the Apsara every limb of whose body was endued with great beauty, the ascetic addressed her, saying, 'O thou of slender waist, I have a doubt in my mind. Do thou explain it.'

Bhishma continued, 'Thus addressed by the Rishi, the Apsara said unto him, 'If the subject is one which is known to me and if thou thinkest me competent to speak on it, I shall certainly say what is in my mind.'

Narada said, 'O amiable one, I shall not certainly appoint thee to any task that is beyond thy competence. O thou of beautiful face, I wish to hear from thee of the disposition of women.'

Bhishma continued, 'Hearing these words of the celestial Rishi, that foremost of Apsaras replied unto him, saying, 'I am unable, being myself a woman, to speak ill of women. Thou knowest what women are and with what nature they are endued. It behoveth thee not, O celestial Rishi, to set me to such a task.' Unto her the celestial Rishi said, 'It is very true, O thou of slender waist! One incurs fault by speaking what is untrue. In saying, however, what is true, there can be no fault.' Thus addressed by him, the Apsara Panchachuda of sweet smiles consented to answer Narada's question. She then addressed herself to mention what the true and eternal faults of women are!'

Panchachuda said, 'Even if high-born and endued with beauty and possessed of protectors, women wish to transgress the restraints assigned to them. This fault truly stains them, O Narada! There is nothing else that is more sinful than women. Verily, women, are the root of all faults. That is, certainly known to thee, O Narada! Women, even when possessed of husbands having fame and wealth, of handsome features and completely obedient to them, are prepared to disregard them if they get the opportunity. This, O puissant one, is a sinful disposition with us women that, casting off modesty, we cultivate the companionship of men of sinful habits and intentions. Women betray a liking for those men who court them, who approach their presence, and who respectfully serve them to even a slight extent. Through want of solicitation by persons of the other sex, or fear of relatives, women, who are naturally impatient of all restraints, do not transgress those that have been ordained for them, and remain by the side of their husbands. There is none whom they are incapable of admitting to their favours. They never take into consideration the age of the person they are prepared to favour. Ugly or handsome, if only the person happens to belong to the opposite sex, women are ready to enjoy his companionship. That women remain faithful to their lords is due not to their fear of sin, nor to compassion, nor to wealth, nor to the affection that springs up in their hearts for kinsmen and children. Women living in the bosom of respectable families envy the condition of those members of their sex that are young and well-adorned with jewels and gems and that lead a free life. Even those women that are loved by their husbands and treated with great respect, are seen to bestow their favours upon men that are hump-backed, that are blind, that are idiots, or that are dwarfs. Women may be seen to like the companionship of even those men that are destitute of the power of locomotion or those men that are endued with great ugliness of features. O great Rishi, there is no man in this world whom women may regard as unfit for companionship. Through inability to obtain persons of the opposite sex, or fear of relatives, or fear of death and imprisonment, women remain, of themselves, within the restraints prescribed for them. They are exceedingly restless, for they always hanker after new companions. In consequence of their nature being unintelligible, they are incapable of being kept in obedience by affectionate treatment. Their disposition is such that they are incapable of being restrained when bent upon transgression. Verily, women are like the words

341

uttered by the wise. 1 Fire is never satiated with fuel. Ocean can never be filled with the waters that rivers bring unto him. The Destroyer is never satiated with slaying even all living creatures. Similarly, women are never satiated with men. This, O celestial Rishi is another mystery connected with women. As soon as they see a man of handsome and charming features, unfailing signs of desire appear on their persons. They never show sufficient regard for even such husbands as accomplish all their wishes, as always do what is agreeable to them and as protect them from want and danger. Women never regard so highly even articles of enjoyment in abundance or ornaments or other possessions of an agreeable kind as they do the companionship of persons of the opposite sex. The destroyer, the deity of wind, death, the nether legions, the equine mouth that roves through the ocean, vomiting ceaseless flames of fire, the sharpness of the razor, virulent poison, the snake, and Fire--all these exist in a state of union in women. That eternal Brahman whence the five great elements have sprung into existence, whence the Creator Brahma hath ordained the universe, and whence, indeed, men have sprung, verily from the same eternal source have women sprung into existence. At that time, again, O Narada, when women were created, these faults that I have enumerated were planted in them!'"

There is another Hindu sacred text called Shatapatha Brahmana. It describes details of Vedic rituals, including philosophical and mythological background to basically teach the mankind. And what exactly it teaches…

Shatapatha Brahmana 14:1:1:31

And whilst not coming into contact with Sûdras and remains of food; for this Gharma is he that shines yonder, and he is excellence, truth, and light; but woman, the Sûdra, the dog, and the black bird (the crow), are untruth: he should not look at these, lest he should mingle excellence and sin, light and darkness, truth and untruth.

Again in Rig Veda 10:XCV:15

"Nay, do not die, Pururavas, nor vanish: let not the evil-omened wolves devour thee. With women there can be no lasting friendship: hearts of hyenas are the hearts of women."

Various anomalous brain activities of men throughout the world gave rise to the religious doctrines of the ancient times. And their innate desire to have power over everything including the very existence of women became an undeniable part of the man-made religious doctrines.

And the torturous indoctrination goes on in modern times by the Pastors, Imaams, Rabbis, Brahmins and Swamis. Religious backgrounds may vary but the ingrained human elements of power, control and desire never change no matter what the religion.

In a transcendental state of cerebral activity a handful of male savages from the Bronze Age propagated the doctrines of religions influenced by their deepest instincts. Whether they were written by the ancient Indian sages or by the Nomadic individuals, basically all the religious regulations in the ancient texts were simple representation of human desires and instincts expressed during anomalous temporal lobe activity. Good or bad, all are on humans, not some Supreme Sentient Being up in the sky.

And all the harmful elements of religiosity and spirituality become the point of attack for the naïve over-reasonable atheists. It is true that atheism is based on reason and logic that are the basic elements of science. Like you read earlier that spiritual beliefs are hardwired in the cognitive structure, so any kind of atheistic attack of logic would only make the beliefs stronger.

Someone recently asked me *"Do you think that there is any possibility of the existence of God?"*

My answer to this question was and still is *"I don't know"*. And the fact is nobody knows. Religious people believe in the existence of God because it brings a kind of comfort and bliss, and cultural influence matters as well. Therefore, the believers believe in the existence of a Supernatural Being, but nobody knows whether that entity actually exists.

So, in this case what matters the most is the bliss that comes along with religiosity and spirituality. And today we can simulate the inexplicable bliss inside the laboratory without giving a phone call to the Lord Almighty. Now the atheists would think that what about the horrifying violence caused by various religious groups! To the atheists, I would like to show something. Any kind of experience has both good and bad sides. Science itself today is used in various harmful activities.

Radioactivity has the potential to power a city, and at the same time it has the potential to become a weapon of mass destruction. That's exactly what happens to religion and spirituality. No matter how good you are, there will always be someone around the corner to do harm to the society in the name of religion. Atheism is justified and even beneficial to some extent as long as it doesn't take away a person's hope. Hope is something that keeps us humans alive. It's scientifically a part of our brain functions that developed through millions of years, so that we could become better than our primitive ancestors.

Let me talk a little straight with you. Just imagine yourself to be a peaceful, loving and good human being with Islamic beliefs (not an extremist). You are born and brought up in a loving Muslim family. Since the very first day that you gained consciousness, you have been waking up to the mysterious yet somehow surreal melody of Azan.

Now let me ask you something. How would you feel, when every time there is a terrorist attack somewhere in the world in the name of Allah and Jihad, the blame always comes upon you, just because you are born in a Muslim family!

It's really very simple. Faith and hope remove worry, anxiety, and fear. Human life becomes very painful and burdensome if a person has no one to trust and love. Then why should it bother an atheist, if a mother who just has lost her child, takes up a doll of baby Jesus or Krishna and pampers it like her own child, while in the process she actually succeeds in coping with her traumatic situation!

We can say that Atheism has two categories – implicit atheism and explicit atheism. Implicit atheists are those who just don't possess any specific religious belief. They just want to sleep on Sundays. And in most cases they just prefer being humans, and nothing else matters.

And explicit atheists are those who possess conscious and extreme denial of religious beliefs. In many aspects extreme religiosity and explicit atheism both can become harmful to people mentally and physically. For example Richard Dawkins has recently suggested that children should not be told fairytales. He described fairytales as pernicious. While our very own Albert Einstein once said "*If you want your children to be intelligent, read them fairy tales. If you want them to be more intelligent, read them more fairy tales.*"

In early childhood, fairytales actually stimulate the brain's creativity. And this creativity never goes away. The thing is, when a kid grows up into an adult, he or she would naturally shake off unnecessary absurd childish beliefs in those wonder tales, but the imaginative essence of them stays deep-rooted in the mind throughout the lifetime. So, forget religion, it is technically inhuman to take away a child's fascinating childhood. Dawkins is a remarkable biologist and nowadays a leading figure in the atheist community with an amazingly reasonable mind, but sometimes he is a little too harsh about the scientific reality. Especially when he says that all religious people are delusional, it is the most unscientific and downright baseless thing to say for a scientist. Beliefs are personal domain. And conflict rises in the society when one person tries to impose his personal beliefs onto another person, no matter what the belief. From this psychological perspective, there is really not much difference between religious extremists and explicit atheists. All of them just want to impose their personal ideas and beliefs onto the entire human population.

I believe humanism is the only thing that matters globally regardless of religion, caste and creed. Being human itself means, one gives preference to humans. So, if a person is happy in his or her personal life with certain religious or spiritual beliefs and not offending anyone with them, who am I to say he is Delusional. It is one thing to be rational and another to be so ruthlessly rational that one becomes completely blind to all the positivity of religiosity and spirituality. If you look close enough into the functioning of the human mind, it would be impossible for you to not understand the evolutionary importance of religious or spiritual beliefs in most people.

The 'existence' and 'non-existence' are attributes of the human mind. The Reality is beyond both. I as a scientist can say that we can only pursue understanding the nature of the human mind. Each answer will give rise to another question.

"Out beyond ideas of wrongdoing and rightdoing there is a field. I'll meet you there"

- *Rumi*

* * *

BOOK V

LOVE SUTRA

THE NEUROSCIENTIFIC MANUAL OF LOVE

INTRODUCTION

Love is patient and kind. Love is not jealous or boastful or proud or rude. It does not demand its own way. It is not irritable, and it keeps no record of being wronged. It does not rejoice about injustice but rejoices whenever the truth wins out. Love never gives up, never loses faith, is always hopeful, and endures through every circumstance ... love will last forever!

In an attempt to embark on this scientifically sensual journey of love, what other lines could be better than the above ones from the Corinthians! This little passage itself is a peerless descriptor of love. To be absolutely honest, I couldn't find any more perfect verse than this. It's just the ideal illustration of Love everyone seeks. It doesn't get any more ideal than this.

This one word "Love" has revolutionized the whole planet in an evolutionary way. Without this, none of us might have been born. It is Mother Nature's magical wand by the touch of which everything turns surreal and captivating. Love is the most important ingredient of human life. Without it, life seems bleak. There is no other yearning like the yearning for a beloved one. That's why the tiniest glance at your dearly beloved, can

351

make you experience heaven right here on earth. And to be utterly honest, who wants Jesus, when you can have the warm embrace of the person who is the living embodiment of all your desires. So naturally on Christmas it comes out of your mouth

"All I want for Christmas is you".

The truly fantastic sensation of love has been inspiring artists, scientists, philosophers and thinkers for ages. Albert Einstein said *"any man who can drive safely while kissing a pretty girl is simply not giving the kiss the attention it deserves".* Geniuses around the world came up with various masterpieces under the spell of love. Schrodinger's Wave Equation, Hawking's Hawking Radiation, Tagore's songs are just a few among the plethora of scientific and philosophical literature created under the enigmatic and warm influence of love. So, technically it is totally worth being crazy in love.

"I seem to have loved you in numberless forms, numberless time

In life after life, in age after age, forever."

Rabindranath Tagore

"Would you become a pilgrim on the road of love?

The first condition is that you make yourself humble as dust and ashes."

Rumi

But if the mental state of love is so soothing then why do people often fail to maintain a healthy relationship?

The answer lies deep inside the human mind. Once you observe the fascinating biological structures of a loving male and female mind very closely, you shall discover something rather extraordinary. Human mind is sexually dimorphic. And this bizarre dimorphism is the major cause of all break-ups.

It is in fact very simple. The male and female brains are wired differently. For this reason, the male brain perceives every situation of daily life from a typically male perspective and expects the female to do the same, while the female brain observes everything in a feminine manner and expects the male to do the same. And from this very desynchronization emerges all the predicaments of a relationship.

Therefore, it is the lack of insight into the inner arena of the opposite's as well as one's own mind that ruins a relationship. There are various beautiful neurobiological structures at play behind the marvelous experience of love. And these structures work differently in the male and female brains.

Throughout the lifetime, the entire neurobiology of a human being goes through relentless perplexing transformations. These sexually dimorphic neurobiological changes create a person's personality. These unique makeovers of the male and female biology hold the key to a sustainable romantic relationship. So, let's get going in a journey of befriending the most remarkable neurobiological structures of the human mind that ensured the survival of a species for millions of years. Let's discover the pathway towards a healthy and cheerful relationship.

* * *

CHAPTER 1
FALLING IN LOVE

'Tis better to have loved and lost

Than never to have loved at all.

-    *Alfred Lord Tennyson, In Memoriam*

A life without having loved someone is a life never lived. Anybody who has ever fallen in love can tell without wasting a second, that falling in love is an incomprehensible sensation and perhaps the best feeling ever. All of you can relate to it one way or another! The euphoria, madness and bliss that come along the way of falling for someone are just beyond any human interpretation.

When you open your eyes in the morning and with the first ray of sunshine have a glance at your dearly beloved, it seems as if you can see the radiance of a full moon in broad daylight. Then probably some of you wonder *"how can this be science?"*

355

The very first few days when you actually start having symptoms of falling for someone special, are the days of heavenly bliss and unreasonable madness. This specific "madness" is one of the most rudimentary elements of the foundation of love. Along comes "Euphoria". It feels like you have grown wings and you can fly around without a single care in the world. Everything starts to seem beautiful and better. The sun shines a bit brighter and the birds twitter a little louder and sweeter. As if you are stuck in an enchanting dream. You get butterflies in your stomach whenever the special person casts a blazing gaze upon you. And off course you all know about the thumping of heart and dilation of pupils.

This "Mad Love" is actually the most unpolished form of love that later leads to Romantic Love through the establishment of romantic relationship. Romantic Love is the foam of bonding that is crucial for the survival of the human species. A simple yet truly beautiful explanation of romantic love was given by the great singer Bob Marley,

"Only once in your life, I truly believe, you find someone who can completely turn your world around. You tell them things that you've never shared with another soul and they absorb everything you say and actually want to hear more. You share hopes for the future, dreams that will never come true, goals that were never achieved and the many disappointments life has thrown at you. When something wonderful happens, you can't wait to tell them about it, knowing they will share in your excitement. They are not embarrassed to cry with you when you are hurting or laugh with you when you make a fool of yourself. Never do they hurt your

feelings or make you feel like you are not good enough, but rather they build you up and show you the things about yourself that make you special and even beautiful. There is never any pressure, jealousy or competition but only a quiet calmness when they are around. You can be yourself and not worry about what they will think of you because they love you for who you are. The things that seem insignificant to most people such as a note, song or walk become invaluable treasures kept safe in your heart to cherish forever. Memories of your childhood come back and are so clear and vivid it's like being young again. Colors seem brighter and more brilliant. Laughter seems part of daily life where before it was infrequent or didn't exist at all. A phone call or two during the day helps to get you through a long day's work and always brings a smile to your face. In their presence, there's no need for continuous conversation, but you find you're quite content in just having them nearby. Things that never interested you before become fascinating because you know they are important to this person who is so special to you. You think of this person on every occasion and in everything you do. Simple things bring them to mind like a pale blue sky, gentle wind or even a storm cloud on the horizon. You open your heart knowing that there's a chance it may be broken one day and in opening your heart, you experience a love and joy that you never dreamed possible. You find that being vulnerable is the only way to allow your heart to feel true pleasure that's so real it scares you. You find strength in knowing you have a true friend and possibly a soul mate who will remain loyal to the end. Life seems completely different, exciting and worthwhile. Your only hope and security is in knowing that they are a part of your life."

Love is the most extraordinary feeling that has only recently become a major phenomenon of scientific investigation. With all the findings of neuroscience we finally can glimpse the fascinating picture of all forms of love regulated by fascinating neural mechanisms. And as a whole a deeper understanding of the biological basis of love leads our species towards more rewarding and satisfying romantic relationships.

Mother Nature has provided us with various intricate and complex brain mechanisms just to ensure the cheerful survival of the human species. We even have mechanisms inside our head for the purpose of spirituality and religiosity.

The madness of love is all about flooding of molecules. In the early phase of crazy love your brain is literally flooded with neurochemicals. Hence the pleasure centers of your brain light up like a flashlight. At this stage no criticism can penetrate through the euphoric shield of love and this is exactly what we call *"addicted to love"*.

Love is a natural rewarding state of mind, while addictive drugs are artificial stimuli. However, the brain circuits that are activated when we are in love match those of the drug addict desperately craving for the next fix. Natural rewarding stimulation like love may not be as strong and transcending as that achievable by addictive drugs like LSD, Ecstasy, Pot, Cocaine etc. but unlike them, love has true potential for stress-reduction and health-promotion. Now some might say that Pot reduces stress as well, so what's the difference between Love and Pot! Well, artificial drugs reduce stress temporarily while damaging brain functions in time.

In this context, allow me to quote from my book The
Art of Neuroscience in Everything :

*"Love begins with the stage of primitive lust and attraction. I'm
saying primitive because at this very early stage there is really no
difference between primitive man and modern man. The bodily
characteristics of a person such as, how hot they are, poke the level
of sex hormones (testosterone and estrogen), cortisol and
pheromones. Lust is initiated at this stage through the physical
attraction and flirting. This is an evolutionary behavior of
mankind that biologically enables a human to find a healthy,
fertile and perfect mate.*

*Following the cue of lust, the major attraction symptoms kick in,
which are usually known as the symptoms of love, such as sweaty
palms, tremors in the whole body, restlessness, loss of appetite and
sleep, thumping heart, butterflies in the stomach etc. Such
symptoms occur because the body is flooded with neurochemicals
like Dopamine, Cortisol, Norepinephrine, and Phenylethylamine
(PEA). Once this euphoria wears off, the ultimate and deepest
stage of love prevails that is the attachment phenomenon. And the
chemicals that make this possible are Oxytocin, Vasopressin and
Endorphins. As time goes by, the crazy love sensation diminishes
and the feeling of closeness and attachment grows and prevails till
the last breath of life."*

So, in the beginning of your love life it is the primitive
wild beast within you that drives you towards the
person incognito. No matter how much we want to
deny that unpolished, uncivilized beast within each one
of us, every now and then it gives signs of its innate
existence in every walk of life. That is exactly why some
of you might like the movie 'Fifty Shades of Grey'. And

actually we may resist that beast from getting out and pretend to be decent, but inside we all know how bad we are. That's exactly what Freud called the "Id".

Research has shown that due to this primitive programming it takes the male brain only one fifth of a second to classify a woman as sexually hot or not. The unconscious mind reaches to the conclusion long before a man's conscious mind engages in the process.

For men and women, the initial calculations about romance are totally unconscious, and they're very different. Men are chasers and women are choosers. It's our inheritance from the primitive ancestors who learned, over millions of years, how to propagate their genes. The attraction to an hourglass figure – large breasts, small waist, flat stomach, and full hips, is ingrained in the neurobiology of men across all cultures. This shape tells the male brain that she's a young, healthy and fertile mate.

So, whenever a man ogles at a woman's breasts or hips, in most cases it doesn't mean he intends to sexually assault her. Rather it's an evolutionary instinct of the male brain that cannot be erased completely from the genetic blueprint. In fact, biologically speaking, it's actually quite flattering than offending for a woman.

Darwin noted, males of all species are made for wooing females, and females typically choose among their suitors. This is the brain architecture of love, engineered by the reproductive winners in evolution. Even the shapes, faces, smells, and ages of the mates we choose are influenced by patterns set ages ago.

Falling in love is one of the most irrational behaviors or brain states imaginable for both men and women. The brain becomes "illogical" in the throes of new romance. If we could travel along a person's brain circuits as he or she is falling in love, we'd begin in an area deep at the center of the brain called the ventral tegmental area (VTA). We'd see the cells in this area rapidly producing dopamine.

Dopamine is the brain's feel-good neurotransmitter for motivation and reward. As the brain gets filled with dopamine, the person starts to feel a pleasant buzz. The flood of dopamine stimulates the nucleus accumbens (NAc), the brain region involved in the feeling of pleasure and reward, or simply the brain's reward center.

In a male brain, we'd see the dopamine being mixed with testosterone and vasopressin, while in a female brain, it gets mixed with estrogen and oxytocin. The fusion of dopamine with these other hormones makes an addictive impact over the person, leaving both the male and female exhilarated and head over heels in love.

And the last stand of this mad love is the caudate nucleus (CN), the area for memorizing the look and identity of whoever is giving pleasure. Here we'd see all the minuscule details about the woman or the man being indelibly chiseled into the permanent memory. At this point your beloved one becomes literally unforgettable. Once the train of love has made these three stops at the VTA, NAc and CN, we'd see the

361

brain's lust and love circuits merge together as they focus only on the beloved one.

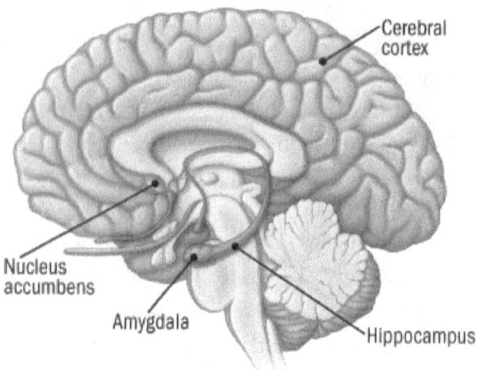

Figure 1.1 Nucleus accumbens, Amygdala, Hippocampus and Cortex

The brain circuits for passionately being in love or the so-called infatuation-love share brain circuits with states of obsession, mania, intoxication, thirst, and hunger. Also, as I mentioned earlier the brain circuits that are activated when we are in love match those of the drug addict desperately craving for the next fix. The most remarkable data on the neurobiological foundation of romantic love come from the studies conducted by Bartels and Zeki (2000), Aron et al. (2005, 2011), Xu et al. (2010), and Ortigue et al. (2007).

The amygdala (fear-alert system) and the prefrontal cortex (judgment and critical thinking system) are turned way down when the love circuits are running at their full potential. This is why we become literally blind to the shortcomings of our dearly beloved. The same thing happens when people take Ecstasy. So romantic

love is a natural way of getting high. The classic
symptoms of early love are also similar to the initial
effects of drugs such as cocaine, heroin and morphine.
Narcotics trigger the brain's reward circuit, causing
effects similar to romance. Hence the well-known
phrase *"addicted to love"* is scientifically quite literal and
accurate. Studies have shown that this early ecstatic
stage of romantic love lasts for around six to eight
months. During this stage break-up can be catastrophic
leading to withdrawal like symptoms, as the body keeps
hankering for the sensation of euphoria connected to
the person. These early months of a relationship,
romantically involved partners literally crave for each
other and feel undeniably dependent on each other.
This is such an extreme state that the partner's well-
being becomes more important than one's own.

* * *

CHAPTER 2
COUPLING OF COUPLES

Life is short, Break the Rules.

Forgive quickly, Kiss slowly.

Love truly. Laugh uncontrollably

And never regret anything

That makes you smile.

- Mark Twain

Attachment is the most beautiful and evolutionarily significant state of a mature romantic relationship. The powerful emotional bond between two romantically involved partners starts to build up right after the euphoria of mad love wears off.

The lessons of relationship that our primordial ancestors learned are deeply encoded in our modern brains as neurological circuits of love. They are present from the moment we're born and activated at puberty by the cocktail of neurochemicals. It's an elegant synchronized system. At first our brain weighs a potential partner, and if the person fits our ancestral wish list, we get a spike in the release of chemicals that makes us dizzy with a rush of unavoidable infatuation. It's the first step down the primeval path of pair-bonding.

In the rose-colored world of pair-bonding the most influential arrows of cupid are Oxytocin and Vasopressin. These two are the most extreme persons of interest in all kinds of love and trust on planet earth. Attachment in any relationship is only possible by the grace of these neurochemicals. And attachment is exactly what keeps a relationship alive and healthy. So, all the so-called philosophical notion of "love without attachment" or "detached love" are biologically non-existent on this planet. We humans are biologically designed through millions of years of evolution to grow attachment. Love cannot survive without attachment. In fact, a mature romantic relationship begins after the early euphoric stage, when a couple starts getting coupled together.

Our primitive ancestors learnt various behavioral characteristics like jealousy, possessiveness and aggression to ensure the survival of their wild love life in the harsh environment of Mother Nature. And all those behavioral responses eventually got engraved in our genetic blueprint. So, these are not the enemies in the path of a healthy relationship, rather when utilized properly they can even kindle the spark in a dying relationship.

But why exactly, such a common biological trait of pair-bonding evolved? There is a simple evolutionary answer to that question.

Evolutionarily speaking, love is all about procreation. In this case, only erotic passionate love making does not guarantee successful procreation. A lot more effort needs to be invested by the parents to actually ensure the survival of their progeny. For us humans, this can be up to twenty years. For this purpose, Mother Nature developed strategies beyond the "one time fling" approach to make mating partners collaborate until their progeny can survive on its own. Hence the neurological circuits of pair-bonding evolved. Along the way it led to the neuropsychological arrival of monogamy as a favored type of relationship. However, scientifically speaking, women tend to be more monogamous than men, whereas the tendency of men is to be polygamous and promiscuous. Among the hundreds of human cultures throughout the world, only one, the Thodas of South India, have officially endorsed polyandry (the practice of having more than one husband or male mate).

For a man, the optimal evolutionary strategy is to disseminate his genes as widely as possible, given his few minutes (or, alas, seconds) of investment in each encounter. It all makes simple evolutionary sense, since a woman invests a good deal of time and effort - a nine month long, risky, strenuous pregnancy, in each offspring. Naturally she has to be very discerning in her choice of sexual partners.

Now the question rises, if men are biologically polygamous and women are monogamous, then how can a romantic relationship ever last for long? The answer is again in the evolution of various brain regions. It is true that men will always be men with their innate wild attraction to all the breasts and hips around even while being in a relationship. But through the process of Darwinian natural selection, an amazing brain region evolved, i.e. the pre-frontal cortex.

Prefrontal cortex is the area of the brain that gives you the ability to keep all your momentary emotional impulses in check. So, even though a man is biologically incapable of stopping his testosterone level to go high when he visualizes a hot lady, he still can choose whether to act upon that momentary impulse of libido.

Just imagine the concern that Mother Nature has for us. She put all her excellence in designing various brain circuits with utmost care so that we could lead a healthy, happy and abundant family life. She programmed us to go crazy with just a glance of our beloved ones.

In various studies we have found that merely having a look at the picture of your dear ones turns on the insula, anterior cingulate cortex, putamen, retrosplenial cortex and caudate like a nuclear furnace. The insula and anterior cingulate cortex are typically associated with emotion oriented attentional states, whereas the retrosplenial cortex is involved in episodic memory recall, imagination, and planning for the future.

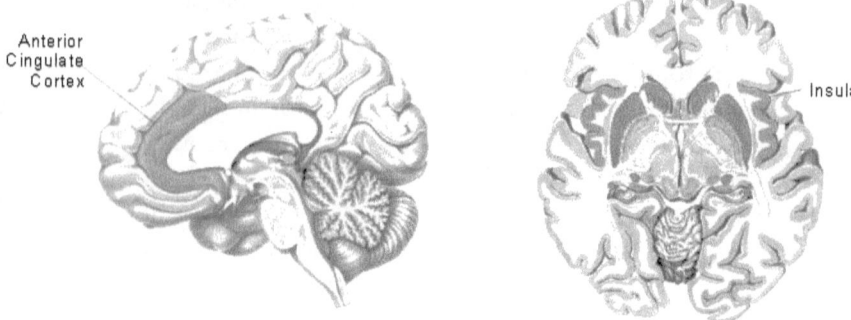

Figure 2.1 Anterior Cingulate Cortex and Insula are shown as darker regions

And off course the brain's love circuits share several brain regions with sexual arousal circuits. In several studies it has been found that certain regions do appear to be consistently heightened in response to sexual stimuli, such as the hypothalamus, putamen, visual cortex, inferior temporal cortex, orbitofrontal cortex, anterior cingulate cortex, parietal cortex, temporo-parietal junction, insula, ventral striatum, anterior temporal areas, amygdala, and basal ganglia (Ferretti et al. 2005, Fonteille & Stoleru 2010, Karama et al. 2002, Maravilla & Yang 2007, Moulier et al. 2006, Redoute et al. 2000, Walter et al. 2008). These studies have also pointed out a few regions that are distinctively active only in response to romantic stimuli. And those regions are caudate and ventral tegmental area. This significantly implies that the brain circuits of love and sexual arousal are anatomically distinctive yet intertwined.

But the craftsmanship of nature does not end just here. Like a nourishing mother, she embedded the ingredients of attachment right inside our head. Those ingredients are Oxytocin and Vasopressin. They play a critical role in forming a concept of our partner whom we want to be with. They appear to build a strong profile of the mating partner through odor. The odor comes to be associated with a pleasurable and rewarding encounter with a particular partner. The same works in the visual domain. Oxytocin is not only responsible for the bonding of couples but also it is involved in maternal love towards a baby, whereas vasopressin is responsible for the commitment of the male towards his mate.

But the brain region activation in women that correlates with maternal love is not identical to the one with romantic love. An interesting distinction lies in the strong activation of fusiform gyrus that is involved in the attention to faces in maternal love. This counts for the importance of reading children's facial expressions to ensure their well-being. This leads to the constant attention that a mother pays to the face of her child. Many of us scientists believe that damage or abnormality in the fusiform gyrus leads to the condition called "prosopagnosia" or simply "face blindness". Another interesting difference is the hypothalamus, which is involved in sexual arousal, thus only in romantic love.

The influence of Oxytocin and Vasopressin is far more delicate than you can imagine. These two incredible hormones go to great length to keep us from being promiscuous. To illustrate this, let me tell you the story of the prairie and the montane voles. It is a story of great biological interest. Among these two species, the prairie voles are mostly monogamous in nature, while the montane voles are promiscuous. Due to their brain circuits, the montane voles cannot maintain a healthy long-term relationship. If the release of Oxytocin and Vasopressin is blocked in prairie voles, they too become promiscuous. If however prairie voles are injected with these hormones but prevented from having sex, they will still continue to be faithful to their partners through a chaste monogamous relationship. That makes me wonder, what if we just inject the montane voles with Oxytocin and Vasopressin! Makes sense right!

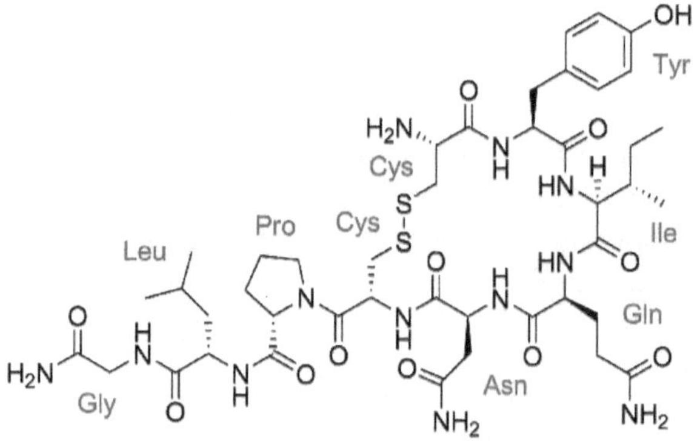

Figure 2.2 Oxytocin - The Hormone of Love, Trust and Bonding

One might think that injecting the montane voles with these two hormones will somehow magically transform them into faithful monogamous creatures. But quite unfortunately it doesn't work that way. An injection of these love potions don't render them monogamous. Once secreted by the pituitary, these neurochemicals can only act if there are receptors for them in the brain. In the prairie voles there is an abundance of receptors

for Oxytocin and Vasopressin in the reward centers of
the brain. While on the contrary in the montane voles,
receptors for these two hormones are not as abundant.
Ergo, injecting the montane voles with excessive
amounts of Oxytocin and Vasopressin doesn't make
them monogamous, since there are not sufficient
receptors for them in the reward centers.

Figure 2.3 Left: Monogamous Prairie Vole, Right: Promiscuous Montane Vole

There is a genetic cause behind this receptor variability.
Prairie voles carry a longer version of the vasopressin
receptor gene which makes them way more
monogamous in behavior than the montane voles. Our
two closest primate cousins, chimpanzees and bonobos
also have different lengths of this gene, which match
their social behaviors. Chimpanzees, who have the
shorter gene, live in territorially based societies
controlled by males who make frequent, fatal war raids
on neighboring troops. While on the other hand,
Bonobos are run by female hierarchies and seal every
social interaction with a bit of sexual impression. They
are exceptionally social and have the long version of the
gene. The human version of the gene is more like the
bonobo gene. Differences in partner commitment may

therefore be related to our individual differences in the length of this gene and in hormones.

So the ongoing joke among the women scientists is that the women should care more about the length of the vasopressin gene in their mates than about the length of anything else. Maybe someday there will be a drugstore test kit, similar to a pregnancy test, to know how long this gene is, so a woman can be sure she's getting the best guy before she commits. Just in a manner of speaking, male monogamy may therefore be somewhat predetermined for each individual and passed down genetically to the next generation. Possibly devoted fathers and faithful partners are born, not made or shaped by a father's example. However, as genetic engineering progresses, perhaps one day a woman will be able to alter the length of the vasopressin receptor gene in her man, to make a perfect faithful partner out of him.

Anyways, the brain circuit for attachment serves the crucial evolutionary purpose of maintaining and promoting the survival of the species. All the brain circuits were carefully crafted by Mother Nature with accurate precision so that bonding becomes a rewarding experience.

* * *

CHAPTER 3

UNDERSTANDING - THE KEY

Let's get straight to the point. It's not love or care that actually keeps a relationship alive. It's "understanding". But the underlying essence of this mysterious term is not as simple as it looks. Any human being who is in a romantic relationship has some kind of acquaintance with this term. But the circumference of "understanding" doesn't just end at sensing the partner's deepest feelings. It transcends way beyond that area. It encompasses an enormous world of mysterious traits of the human mind, which even the possessor of the mind is not conscious of.

How can you understand your partner's confusing behavior when apparently it doesn't make any sense! How exactly can you find meaning in your partner's seemingly meaningless actions!

William Shakespeare said in A Midsummer Night's Dreams *"Love looks not with the eyes, but with the mind, And therefore is winged Cupid painted blind"*. So, in order to possess the key to a healthy, cheerful and long-term relationship we need to dive into the deepest corners of the male and female minds. Let's get wet in the monsoon of "True Love".

In the path of "True Love", the most elementary thing is to pick up emotional cues. You must know whenever your partner is feeling blue, even if he or she is not expressing anything. Sounds like a lot of magical effort right! But actually it's no effort at all. Again, we must thank our brain, that it practically has gifted us with the ability to read the mind of our partner. Whether you call it "gut-feeling" or "intuition", by all means it's technically mind-reading. And biologically speaking, women are better at this than men. There is a fascinating interplay of brain circuits behind your gut-feeling. Gut feelings are not just free-floating emotional states but actual physical sensations that convey meaning to certain areas in the brain. And studies have shown that the areas of the brain that are involved in gut feelings are larger and more sensitive in the female brain.

At first a woman begins receiving emotional signals from another person's facial expressions, hand gestures, body postures and breathing rates through firing of the mirror neurons. There is no mysticism involved in it. It's just beautiful biological design.

Brain-scan studies have shown that the simple act of observing another person in a particular emotional state can automatically trigger similar brain region activity in the observer by the grace of the Mirror Neuron System (MNS), this is what we call "emotional empathy". And females are especially good at this kind of emotional mirroring. After the mirror neurons play their part, the body sends a message to the insula and anterior cingulate cortex (refer to Figure 2.1 in the previous chapter). The insula is an area in a classic part of the brain where gut feelings are first processed. The anterior cingulate cortex, which is larger and more easily activated in females, is a critical area for anticipating, judging, controlling, and integrating negative emotions. A woman's pulse rate suddenly bumps up, a feeling of tension felt in her belly and the brain interprets it as an intense emotion.

So, being able to guess what another person is thinking or feeling is technically very much biological. And overall, between the male and female, the female brain is efficient at assessing the thoughts, beliefs, and intentions of others, based on the smallest hints more quickly than the male brain. Such unique feature in the female brain is yet another product of evolutionary practice, as throughout evolution woman had to be very receptive of the facial expressions of her child, in order to ensure its well-being.

In the male brain, most emotions trigger less gut sensation and more rational thought. The typical male brain reaction to an emotion is to avoid it at all costs and find a rational solution to the problem. I remember one of my colleagues once asked her scientist husband *"Why do men respond to emotional issues with logic instead of feelings?"* He laughed and said, *"The real question is why women don't."*

The mirror neurons of a man allow him to briefly feel the same emotional pain he sees in his woman's face. Next, the temporo-parietal junction activates his brain's analytical circuits to search his entire brain for solutions. This is called "cognitive empathy". The male brain is able to use the temporo-parietal junction starting in late childhood, and after puberty a man's reproductive hormones reinforce the preference for it. Researchers have found that the temporo-parietal junction keeps a firm boundary between emotions of the "self" and the "other". This prevents men's thought processes from being infected by other people's emotional weakness, which strengthens their ability to cognitively and analytically find a solution without being vulnerable.

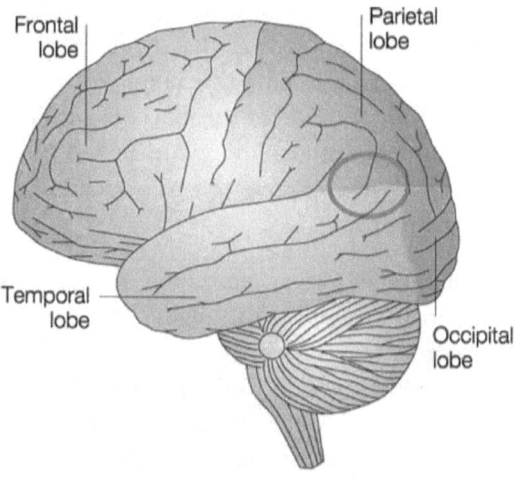

Figure 3.1 All Four Lobes of the Brain, and you can see the temporo-parietal junction indicated in circle

Many women in relationship often complain that their men are blind to the emotional signals they send. That's not actually their fault. While the female brain is a high-performance emotion engine, the male brain is not so skilled at reading facial expressions and emotional innuendoes like signs of despair and hopelessness. Men pick up the subtle signs of sadness in a female face only 40 percent of the time, whereas women can pick up these signs 90 percent of the time. The only way to penetrate their shell of logical thinking, is to burst into tears. It's only when men actually see tears, they realize, that something's wrong. Perhaps that's why women have evolved to cry four times more easily than men by displaying an unambiguous sign of suffering that men can't ignore. Tears in a woman's eyes literally force a man to feel intense pain. A man feels absolutely powerless when he sees his woman in pain. And then the inevitable happens, that is an automatic comforting hug from the man.

However, a relationship doesn't get ruined because of such a vivid biological difference in the brain circuits of men and women. Rather it's about patience. In a typical scenario a male brain needs to go through a longer process to interpret emotional meaning. Most men just don't bother to take the time to figure out the emotion, and they become impatient.

A man may not be so skilled at picking up emotional signals, all he needs to maintain a healthy relationship is to practice the skill of patience. Over time, he'd learn to recognize when his special lady needs a good cry and then he could simply hold her in his arms and be with her until she's done.

One of the beautiful innate qualities of a woman is to be there with her man during emotionally difficult times, which is why she is often baffled by her partner's inability to put up with her sadness or despair. Women are neurologically wired to respond to the distress of other people. So when men say *"women blow things out of proportion"*, what they don't realize is that the male brain is a high-performance logicality machine, while on the other hand the woman's brain is neurologically programmed by Mother Nature to be more sensitive to emotional distress.

Researchers at the University of Michigan have discovered that women use both sides of the brain to respond to emotional experiences, while men use just one side. And in my earlier book I have distinctively illustrated the characteristics of both hemispheres of the brain. The left hemisphere is strongly involved in the sense of self, whereas the right hemisphere is responsible for the awareness of others.

Studies have also shown that the connections between the emotion centers in women are more active and extensive than men. In another study, at Stanford University, volunteers looked at emotional images while having their brains scanned. Nine different brain areas lit up in women, while in men only two lit up. No wonder, a woman always sticks around when her man is hurt or disturbed and would do everything in her capacity to make him feel comfortable.

While on the contrary men tend to avoid contact with emotionally distressed people. Men tend to process their troubles alone and expect women would do the same, which is just the opposite of what women expect. It's all about biological designing. Research has also shown that women remember emotional events such as first dates, vacations and big arguments more vividly and retain them longer than men (Seidlitz and Diener 1998, Canli and colleagues 2002). We'll discuss the reason behind it in a while.

In both the male and female brains, an almond shaped structure located deep within the brain called the amygdala is the emotional coordinating system. From the amygdala emotional impulses go to the hypothalamus (the brain's Homeostasis center). Then the hypothalamus raises the blood pressure, heart rate and breathing, and puts the body into fight-or-flight mode based on the intensity of the emotional impulses.

The amygdala also alerts the cortex (the brain's Intelligence center) which analyses the emotional situation and decides how much attention is required. If the intensity of the emotional impulses is high enough then the conscious brain becomes alert and strong conscious emotional sensation kicks in. Then the prefrontal cortex (the brain's decision-making center) plays its part by determining how to respond to the situation.

Women are way better at recollecting minuscule details of the emotional events of life. One reason for this is their highly sensitive amygdala, which is more easily activated by emotional triggers than in men. The amygdala's neural connections to the rest of the brain put it in a unique position to rapidly respond to sensory input and influence physiological and behavioral responses, as well as to influence memory formation in the adjacent hippocampus.

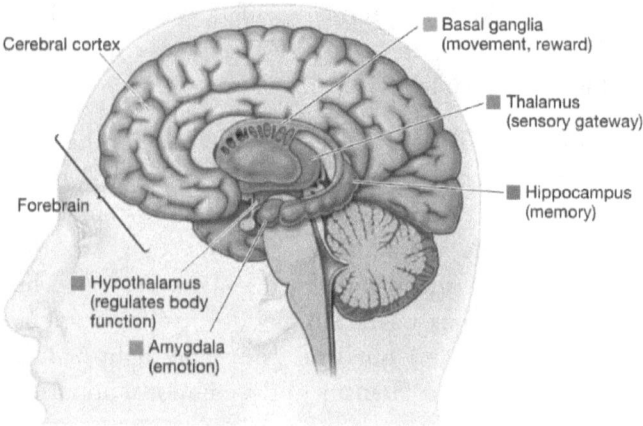

Figure 3.2 Amygdala, Hippocampus, Hypothalamus, Thalamus and Basal ganglia

And one of the most common characteristics of our limbic system is that, the stronger your amygdala respond to an emotional situation, the more details of that situation is indexed by the hippocampus. The hippocampus is the brain's memory formation center. It connects minute emotional senses like smell, sound etc. to memories and send the memories out to the appropriate part of the cerebral hemisphere for long-term storage.

It's really very simple. The more emotional you are in a situation, the more memories you'll have of that situation in the long run. In this context, one thing to mention is that, sexual dimorphism in the hippocampal volume has also been found in many studies. Or in simple terms, the very memory indexer of the brain is actually larger in women than men. From this, you can draw the conclusion yourself.

So, when a man cannot remember the details about the first date, it doesn't at all mean that he does not love his woman any more. It's simply because his brain circuits are unable to retain the information. Men's amygdala and hippocampus work at full throttle in response to any threat to the relationship or any physical danger. In terms of emotional memories, men register memories connected to any kind of threatening situation as vividly as women register all emotional memories. To a man any threat to the relationship can be devastating which he never forgets.

The most innate biological response of men is aggression. It's an irrefutable part of the male existence. As we all probably know that the expression of rage and aggression is greater in men. The cause of men's anger and aggression is their larger amygdala and loads of testosterone receptors in it.

In an adult human brain, the male amygdala is significantly larger than the female amygdala, even when total brain size is taken into consideration (Goldstein and colleagues 2001). While on the contrary women have slightly larger prefrontal cortex and anterior cingulate cortex that are involved in controlling the rage and avoiding any kind of conflict. As a result women have better hold of their anger response than men.

The female brain is engineered to avoid conflicts at all cost, whereas the male brain pleasures conflicts in the purpose of being the boss. Studies have found that though men and women say that they feel anger for an equal number of minutes per day, men get physically aggressive twenty times more often than women. Due to the abundance of testosterone receptors in the amygdala, high testosterone level makes it even more difficult for a man to tame his rage and aggression.

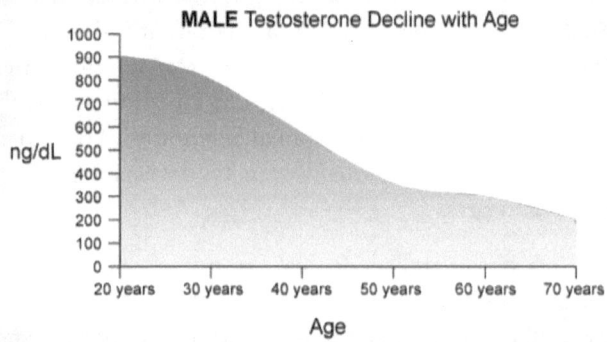

Figure 3.3 Testosterone Level in Men Peaks at Age 20, Thereafter it Declines

Other than testosterone a man's brain circuit for aggression is highly influenced by vasopressin, cortisol and adrenalin. And actually in most cases when a man's anger reaches the boiling point it gives him an utter sensation of pleasure. The pleasure of utter aggression motivates a man to win the fight. All these are evolutionarily encoded inside the male brain to ensure the survival of the progeny at all cost.

Once men enter their fifties or sixties their testosterone and vasopressin levels decline. The ratio of estrogen to testosterone increases with old age. Hormonally speaking, with age the male brain eventually becomes just like the mature female brain. As a result, he doesn't get angry so fast.

* * *

CHAPTER 4
UNPREDICTABLE WOMANHOOD

"Reality" is a beautiful term. It's everywhere around us. Our very existence is predicated on our perception of the reality. This is what we call "subjectivity". Now imagine a reality that is never constant and rapidly changing from week to week. That's exactly what a woman faces in her life.

The hormonal interplay inside a woman's head creates her reality. Her hormones tell her day to day what's important. They mold her desires and values. She goes through these stormy fluctuations of various hormonal states since her girlhood till menopause. It's what in simple terms called "PreMenstrual Syndrome (PMS)".

However, I honestly think it should rather be called "PreMenstrual Storm". And the extreme form of PMS, which only very few women experience, is called "PreMenstrual Dysphoric Disorder (PMDD)". Due to such stormy weather inside a woman's head, her cognitive reality is never as constant as man's. A man's mental universe is like the Himalayas standing millennia after millennia invincible to massive changes in

384

geography, whereas the female universe is like the unpredictable climate change. A man can never even imagine in his wildest dreams how the storm inside the woman's head feels like. These storms often make a woman completely misunderstood by her man. And the common sentences that come out of these misunderstandings go on like this

"You have issues…."

"Nobody can understand a woman…."

But the fact is that these issues are the part of female existence. And a man who doesn't understand his woman during her utter mental turbulence doesn't deserve any woman. There lies another crucial key to a healthy relationship. And if a woman has PMDD then it requires her man to be even more patient and understanding.

If for a second I think of myself as a man in a relationship rather than a scientist, then I know what kind of impact a woman's mental disturbances can have over a relationship. My own girlfriend has the rare extreme form of PMS that is PMDD, which leads to the worst of hormonal mood swings. Every month during these days she turns into a completely different person filled with hopelessness and gloom. As soon as the tides change she comes back to her real cheerful self. And it is her condition that inspired me to write this book in the first place.

In all menstruating women, the female brain goes through mind-blowing changes. Things get really rusty at times, but for most women the changes are manageable. Most weeks of the month women are brainy, creative, enthusiastic, cheerful and optimistic, but a mere shift in the hormonal flood on certain days makes them absolutely hopeless about the future. On those days they tend to hate themselves and their lives. And the most fascinating thing about those days is that the hopelessness caused by hormonal imbalances feels so damn real to a woman that she literally perceives it as the everlasting reality of her life. The hormonal turbulence completely transforms a woman's cognitive reality from a cheerful one to a gloomy one. She becomes absolutely blind to all the cheerful moments of her life. Some women avoid talking to someone because they feel so much restless that they might just explode into tears or bite someone's head off. In this situation the best thing for the man to do is to do nothing and just be there with his woman. And off course she can get really cranky at times so, her man needs to remain patient and unaffected by her words. Once the hormonal storm wears off, she'd come back to her original sunny state.

The voyage of a woman's early life of girlhood is mostly smooth with lots of drama. The temperamental roller coaster ride begins with the onset of menstrual cycle right after the juvenile period of girlhood passes. Here comes the crucial function of the Hypothalamic-Pituitary-Ovarian Axis (HPO Axis). This is the system that drives the roller coaster ride. It is technically a ride of the estrogen-progesterone waves. Testosterone and vasopressin drive a man's manhood whereas a woman's womanhood is guided by estrogen and progesterone. And the HPO Axis keeps strong hold of a woman the whole of her menstruating life.

A healthy menstruation of a healthy woman involves the complex yet synchronized interaction of the hypothalamus, pituitary and ovaries. Menstruation is a natural phenomenon indicating a woman's fertility. Having regular menstrual cycles is a sign that important parts of the female body are working normally. It also prepares the body for pregnancy each month. Since the beginning of life on planet earth, among males and females only the females are endowed with the gift of producing new life. But for this huge privilege a woman has to pay the price almost half of her life by going through the cranky mood swings every single month.

Here is an excerpt from one of my research publications for you to have a look at the functioning of the menstrual cycle :

"The median menstrual cycle length is 28 + 3 days and the average duration of menstrual flow is 5 + 2 days. The cycle, which can be divided into a follicular phase and a luteal phase, results from complex interactions between the hypothalamus, pituitary, and ovary.

….

The proliferative phase, also referred to as the estrogen phase, begins approximately 5 days after menstruation and lasts for about 11 days. E secreted by the ovary stimulates the growth of the endometrium. The stroma cells and epithelial cells begin to proliferate rapidly, uterine glands begin to grow and elongate, and the spiral arteries begin to grow in order to supply the thickened endometrium. Rising E levels then trigger the midcycle LH surge, which induces ovulation. When ovulation occurs, the endometrium is approximately 3-4 mm thick. At this time the endometrial glands secrete a thin, stringy mucus, which protects and leads the sperm into the uterus.

The luteal or secretory phase, also called the progesterone phase, occurs after ovulation and lasts for about 12 days. The corpus luteum secretes high quantities of P and some E. The E causes slight cellular proliferation in the endometrium. P causes significant swelling of the endometrium and converts it to an actively secreting tissue. P also inhibits myometrial (uterine smooth muscle) contractions, in large part by opposing the stimulatory actions of E and prostaglandins. The endometrium reaches a thickness of 5-6 mm about one week after ovulation. The purpose of this process is to prepare the uterus for implantation of the ovum if fertilization occurs.

In the premenstruation or ischemic phase, if pregnancy has not occurred, the coiled arteries constrict and the endometrium becomes anemic and shrinks a day or two before menstruation. The corpus luteum of the ovary begins involution. This lasts about 2 days and is terminated by the opening up of constricted arteries, the breaking off of small patches of endometrium, and the beginning of menstruation with the flow of menstrual fluid.

The desquamation of the endometrium, or menstruation, is caused by the sudden fall in blood P and E, which results from regression of the corpus luteum. This deprives the highly developed endometrial lining of its hormonal support. The immediate result is profound constriction of the uterine blood vessels, which leads to diminished supply of oxygen and nutrients. After the initial period *of vascular constriction, the endometrial arterioles dilate, resulting in hemorrhage through the weakened capillary walls. The menstrual flow consists of this blood mixed with the functional layer of the endometrium. Prostaglandins are thought to mediate both the initial vasoconstriction as well as the uterine contractions accompanying menstrual flow. "*

I know it's a little technical, but the overall idea of the above excerpt is that the tides of female hormones like Estrogen and Progesterone prepare a woman's body for pregnancy every month. And the collateral effect on the woman's mental state is the occasional despair. But it's not just that. The rising tides of estrogen and progesterone mold the female brain circuits into their true feminine form. The blossoming of the feminine hormones makes a woman's brain a high-performance emotion engine. As a result women develops extraordinary power of stress response and along with it comes sharpened critical thinking. And this superpower of stress response continues until after menopause.

After the passing of girlhood a woman's biological instincts guide her to be fantastically sensitive to emotional nuances. In the wrong days of her cycle her man needs to be very cautious of what he says to her. Because, even a slight tinge of pun would sound like loathing to her in those days. Even if he does not intend anything bad, her brain circuits might interpret his comments as highly unpleasant due to the hormonal surges. At that point of time it is absolutely futile to talk about right and wrong in front of a woman, because by the influence of her hormones all she can think of is how much she hates herself, her life and the whole reality she lives in.

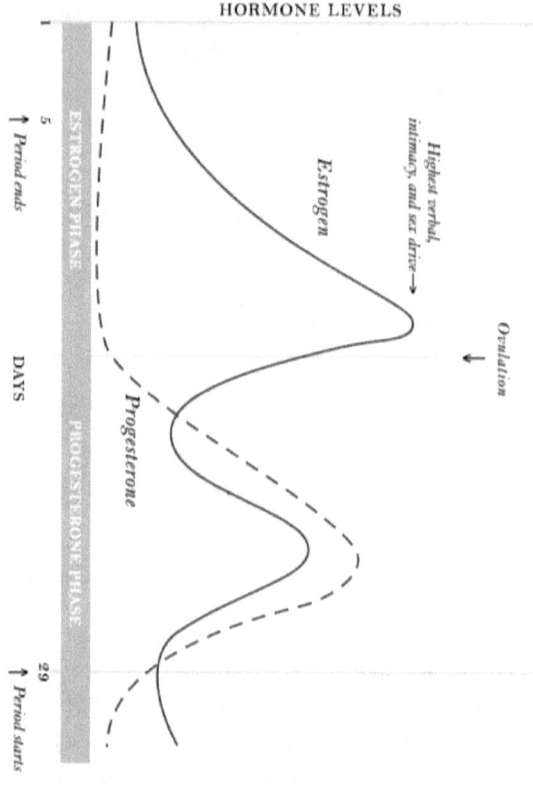

Figure 4.1 The monthly roller-coaster ride of Estrogen and Progesterone waves

* * *

CHAPTER 5
BETWEEN THE SHEETS

Now comes the kinky part of a relationship. It's when things can get really naughty and spicy. But in this fabulous world of sexual bliss where there occurs the communion of two minds through the unification of two bodies, there also arise confusions that often endanger a happy relationship. So, sit tight while I open up to you the mysterious circuits of male and female brains below the belt.

Sexual thoughts float through a man's brain many times a day, while on the contrary a woman has them only one to four times a day. We may have developed a civilized society on this planet, but the primitive mind of humans shall always have imprints of our ancient wild instincts no matter how smart we become. One thing to mention in this context – many philosophers with their metaphysical vanity say that love and lust are two very different things. They consider love to be something sacred, while lust is sinful. But you know what? That's the stupidest thing to say in this century.

The neurological fact is that the brain circuits of lust and sexual arousal are intertwined with the brain circuits of love. In fact without the preliminary unconscious lust, no love can ever even set off. So, regardless of all our pretenses, deep within, we are still unconsciously the same old cave-people. And there is nothing wrong in accepting our innate biological instincts. On the contrary, problems occur when we do not accept our real selves. However, being aware of our deepest biological instincts gives us an evolutionary advantage of programming our own responses and behaviors for a better outcome in our daily life.

The best illustration of such primitive imprints in brain circuits can be found in men. Brain scan studies have shown that even a neutral scenario of a conversation between a man and a woman triggers sexual regions in the observer male brain giving rise to the thoughts of potential sexual rendezvous. Whereas a woman only perceives the scenario as it is — just a simple communique between two people.

But this doesn't mean that men are more sexual than women. It simply means that the male and female brain circuits below the belt are distinctively special in their own way. And the clearest signs of such dissimilarities can be seen in the time required for orgasm. On average it takes a woman three to ten times longer than a man to reach orgasm. There is a biological reason behind it. After all, the evolutionary purpose of coitus is to make babies. So, when a woman reaches climax after her man has already ejaculated, it is more likely for her to conceive. However, orgasm is not compulsory for a woman to get pregnant, but it definitely helps.

Since we are talking about orgasm, let's explore the erotic world of female orgasm. Clitoris is the queen of this erogenous domain. The tip of the clitoris which is called "glans" has more than 8000 sensory nerve endings which communicate directly with the pleasure center of a woman's brain. So, whenever those nerves along with the surrounding vaginal tissues are stimulated by the passion of a man or a vibrator, they evoke electrochemical activity. And once these impulses reach the threshold, they trigger the torrents of feel-good and bonding neurochemicals such as oxytocin, dopamine and endorphins.

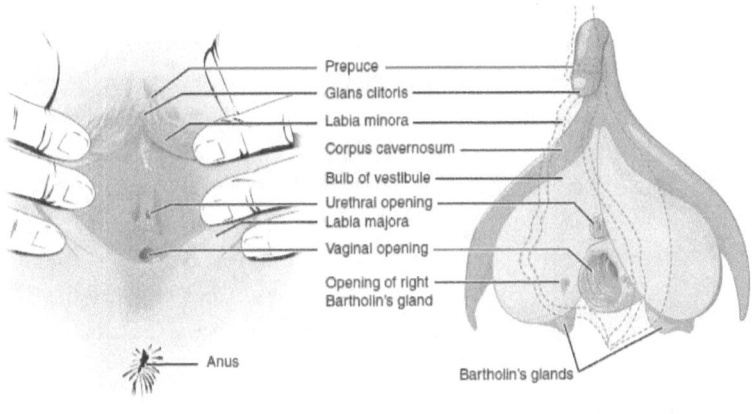

Prepuce

Glans clitoris

Labia minora

Corpus cavernosum

Bulb of vestibule

Urethral opening

Labia majora

Vaginal opening

Opening of right
Bartholin's gland

Anus

Bartholin's glands

Vulva: External anterior view

Vulva: Internal anteriolateral view

Figure 5.1 Vaginal Anatomy

To talk about orgasmic stimulation, I must clear the air about a classic misperception. It's the eternal battle between "vaginal orgasm" and "clitoral orgasm". It all began with the flawed thoughts of Sigmund Freud. He may have illuminated the world of psychoanalysis to a great extent, but in many cases his ideas actually were filled with flaws. In simple terms, he thought that all psychological problems are born out of sexual conflicts. As a modern day neuroscientist, let me clarify something. Physiology and psychology are not at all separate from each other. Rather they are deeply intertwined.

Freud had no clear perception about the anatomy of the human brain, neither did he have flawless notion of the anatomy of clitoris. But, his ideas had such intuitive appeal that many of the words he used, infiltrated popular parlance, although no one thinks of them as science because he never did any experiments. To be honest, alongside his contribution to the psychoanalysis community, many of his theories actually ruined the sex life of many women. For nearly a century his theories of coitus made women believe that they were not quite real women if they only had clitoral orgasm. Over time neuroscientists have discovered that the clitoris is the major organ of orgasm in the female body. And the surrounding area of the vaginal opening assists in reaching climax.

So, the bottom-line is, there is no such thing as "vaginal orgasm" vs. "clitoral orgasm". The entire ring of tissues that surrounds the vaginal opening is connected to the clitoris by nerves and blood vessels. Ultimately all these tissues together are responsible for the female orgasm. This entire erogenous zone is often referred to as the "ring of fire".

I almost forgot to tell you something really interesting about the clitoris from the animal kingdom.

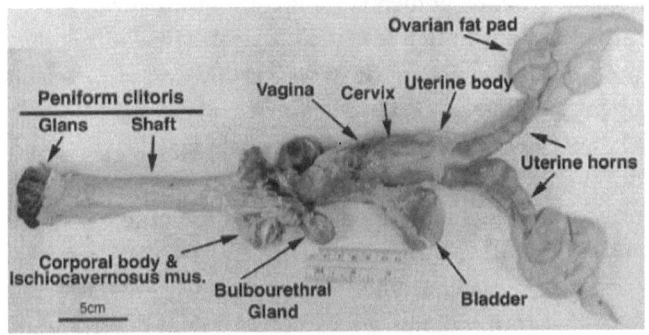

Figure 5.2 Well-Developed Clitoris of Spotted Hyena

Figure 5.3 Pregnant Spotted Hyena with its well-developed versatile Clitoris

The spotted hyena uses its remarkably well-developed and versatile clitoris to pee, mate and give birth.

Now, once the ring of fire has done its job by fulfilling a woman's orgasmic pleasure, her brain is engulfed with dopamine and oxytocin which make her want to cuddle, while inside the male brain, the story is completely different. Many women complain that their men tend to fall asleep right after sex. Well, that is actually true, but the reason is not what women think. This specific phenomenon is called "post-coital narcolepsy". Oxytocin is released in both the male and female brain during and after sex promoting a warm sensation in the body. But while in women it induces the urge to cuddle and talk, inside the men's head post-coital burst of oxytocin triggers the sleep center inside the hypothalamus. So, it is not at all that a man doesn't want to stay awake and cuddle for some time after sex. It is his biology that automatically induces him to sleep. However, we don't yet have the answer to why does Oxytocin work as a sedative only in men and not women?

For a woman it takes a few extra steps to warm up for a sexual encounter. Unlike the male brain, the female brain requires much more than just a hot looking man to get aroused. Her entire biology must be relaxed and cozy in order to have a go into the erotic domain of sexual bliss. Specifically, a hot bath, a good foot rub, chocolates and flattering words before sex unplug the woman from her daily stress and make her ready for the sexual ecstasy. And off course, sexual ecstasy is contagious. So, when the woman is swept off her feet in bed, her man automatically feels the sensuality bursting within his veins. As if some kind of unknown and mysterious force field overwhelms them for limitless eternity.

Sexual bliss is such a state of your mind, that there is no place for anxiety, fear or analytical thinking. So, in order to get a burning turn on, the brainy regions of the brain must be turned off. Researchers have found that it only takes five minutes of casual conversation with a sexy lady for a man's testosterone level to go sky high. And testosterone is the prime driver of the sexual arousal circuits in the male brain.

Figure 5.4 Chocolate is a great ingredient of foreplay. It contains a compound called Phenylethylamine (PEA), that promotes the brain chemistry for love

All it takes for a man to have a full erection is visualization. If a man perceives the potential for any kind of sexual reward in his field of vision his brain sends signal through his spinal cord to the penis. Thereafter blood rushes to the penis making it erect. Here is a beautiful contrast between sensitivity of the tip of the penis and the clitoris. The 8000 nerve endings of the clitoris make the tip of it (glans) more and more sensitive as the woman gets excited, but the penile sensitivity has a different story. Researchers at McGill University have discovered that the tip of the penis (glans) gets less and less sensitive as a man gets more and more aroused.

As blood rushes to the responsible crucial parts of the body such as the penis, clitoris and breasts, we'd discover, brain's fear and anxiety center Amygdala that can interfere in the blissful communion slowly goes dark. Then as the penile thrust begins, the Nucleus Accumbens (the pleasure center of the brain) in both sexes, lights up like a flash light. Symptoms that go along with this heavenly experience such as sensual moaning, quickened breath, trembling, muscle spasms and throbbing heart make you feel on top of this world. And upon reaching orgasm, a tsunami of oxytocin rushes in both the male and female bodies. Right after reaching climax, a glow of contentment radiates from a woman's glistening skin, as the post-coital rush of oxytocin makes her chest and face to blush by expanding the blood vessels. But one clumsy move during sex by the man, and the female amygdala would be back into action, ruining all the sexual interest. Ergo, orgasm goes out of the window.

In our understanding of the true purpose of female orgasm we are way behind the male orgasm. The male orgasm is pretty straightforward. There is no two way about it. But there is lot of hot debate in the scientific community about the purpose of the female orgasm. Alongside the cold perception that there is no actual purpose at all, we do have several hypotheses that give

us some ideas to what might be the purpose of the
female orgasm. So far the major hypotheses on this
matter are: pair-bonding, mate-choice, enhanced fertility
and by-product hypothesis.

The by-product hypothesis is the coldest among all of
them and doesn't hold much water. It states that female
orgasm has no evolutionary function, existing only
because women share some early ontogeny with men.

The pair-bonding hypothesis suggests that the orgasm is
a way of communication between the satisfaction of the
female and male body. Hence it builds intimacy in a
relationship.

Along comes the mate-choice hypothesis. It airs the idea
that the female orgasm inspires a woman's unconscious
mind to cherry-pick the best genetic bet for her scions.
In simple terms, the man who gives a woman intense
orgasms, naturally becomes her unconscious choice.

But biologically the most evident hypothesis is the
enhanced fertility hypothesis. There have been several
studies published on this hypothesis. During orgasm the
uterine contraction sucks up the sperm through the
cervical mucus barrier. In some rare cases the uterine
suction had been so strong that it literally pulled off the
man's condom. Research has shown that when a
woman achieves orgasm any time between one minute
before and forty-five minutes after her man ejaculates,
she pulls significantly more sperm than orgasm-less
coitus. So, technically that's the reason why men are
biologically designed to ejaculate way earlier than
women. It's all about the survival of a species through

procreation. This was the first priority of Mother Nature while she was designing the humans so craftily over millions of years. Ergo, she made the experience of sexual communion the most ecstatic experience on planet earth. It feels so damn good, that we just want to do it over and over again.

A successful orgasm in a woman involves a lot of psychological factors as well, which often get in the way of having a healthy sex life. As I have already mentioned that the female brain is a high-performance emotion engine that retains emotional memories both good and bad much longer than the male brain. Past experiences of a woman's life impose a deep impact over her mental state. And this can interfere in her sexual ecstasy, making it hard for her to have a blissful orgasm.

For example, if a woman had been sexually abused in her childhood or if in her past sexual relationship she had been forced to do something against her will, then such traumatic experience can lead to an unsatisfied sexual life. So, in this kind of circumstances things must be handled with utmost care and concern. Often a combination of sex therapy and trauma therapy proves very effective in such situations. But actually it is not always necessary to visit a therapist, unless the condition is severe. A very simple trick of psychotherapy would come in handy in these cases.

The very rudimentary element of trauma therapy is to vent all the emotions connected to that experience. Once all the repressed and unshared junk is thrown outside the mind, the woman would feel much better.

So, once the foundation of trust in a relationship becomes strong enough, it is much healthier to share all your past memories with each other. Keeping secrets is a kind of emotional repression that actually requires some engagement of the conscious mind in both the man and woman. And over time as the burden of repression gets heavier, it'd start to affect a relationship in a very bad way. In this case, the bottom-line is *"Venting is Healthy"*.

Now let's have a look at some serious psychological factors involved in men's performance in bed. The first thing that many of the women wonder, is *"what is it with men and blowjob!"* Does this mean, men like their package inside women's mouth more than they like it inside their vagina? Technically the answer would be "Yes". And there is a simple biological reason behind it. The sensitive part of a penis works in a different way than the tip of the clitoris. In women, during penetrative sex, the "ring of fire" (clitoris and the surrounding vaginal tissues) slowly gets more and more warmed up sending jolts of electrochemical signals to the brain's pleasure center. But in men, the glans of the penis gets less and less sensitive during penetration to reduce the pain of intercourse.

And Mother Nature's bizarre naughtiness is seen when a woman's soft lips, wet tongue and fingers slowly start to caress a man's penile glans. It sends him into ecstasy as he receives a heightened sensitivity of the glans. Oral sex increases the sensitivity of the tip of the penis in such a way that doesn't occur inside the vagina.

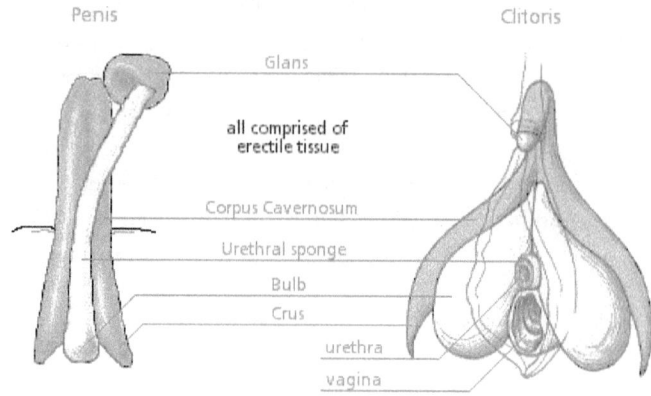

Figure 5.5 Comparative Anatomy of Penis and Vagina

Now let's shed some light over some of the classic male worries concerning sex. Everybody knows about the male concern about size and along with that there is one more worry, which is often called "performance anxiety". In the beginning of sex life, many men worry about their performance in bed. They are concerned about their sexual impression way more than women are aware of. Whether they'll get fully erect! What if they come way too soon! These thoughts frequently float around inside a man's head right before sex. Men need to be relaxed just like women in order to have a proper high performance erection. And daily physical workout can increase men's sexual efficacy. Ergo, maintaining a healthy physique is one of the means to have an excellent love life with bed-bursting sex.

Along comes the size concern. A lot of men wish they had a bigger penis. Research has shown that the average erect penis size ranges from 5.5 to 6.2 inches, which is more than enough for a satisfactory coitus. However, quite surprisingly many men even go through surgery in order to add a few centimeters to their package. The fact is, women are not particularly bothered by the size of the penis. If anything, then it's about how long a man can sustain his erection. And the manly physical features that can turn on a woman are body symmetry, muscles, jawline, smile etc.

At an unconscious level, women are evolutionarily attracted to men who elicit sexual maturity, confidence and sociability. And in case of choosing a mate to raise a family, research has shown that facial hair in men actually attracts the mother in a woman who is planning for babies. The unconscious mind of a woman choses a man with more facial hair as the father of her progeny. However, there is no specific scale for physical attractiveness. It can only be perceived, not measured. And how we perceive attractiveness of the opposite sexes depends on our brain circuits.

* * *

CHAPTER 6

INSTINCTUAL PARENTHOOD

The evolutionary purpose of a relationship is to procreate. Once we fall in love and get committed to a relationship, our biology takes over. And upon the arrival of parenthood, a romantic couple automatically becomes responsible, protective and concerned parents. Mother Nature encoded the instruction manual of parenthood in our genome in order to ensure the survival of the species. As a result, even a couple with no prior interest in kids feels like they are born to be parents as soon as they embrace their newborn. You may be completely unaware of it, but when the time comes your biology guides you through the whole process of parenthood.

In the domain of creating new life, the feminine is the second in command. And off course, as Mother Nature is the alpha of this domain, she programmed the female brain in such a way that it gets saturated with motherly instincts way before the arrival of paternal instincts in men. For this specific purpose she designed an extremely accurate Biological Clock inside human biology. And the most fascinating fact about this clock is that its timing is different in male and female. In female brain circuits, the biological clock has an interesting point of time which gives rise to a kind of craving for a baby. This is what some of us neuroscientists call "baby lust". This lust hits a woman way before she conceives. And again the same old bonding hormone comes into play in this lust. Can you guess which one! Yes, you got it right. It's the hormone of love and trust, Oxytocin. Oxytocin is just as responsible for maternal bonding as it is for romantic bonding. And added to that, if somehow a woman at such a sensitive mental state manages to get a sniff of a newborn's head, then she'd immediately go crazy with the urge to have her own kids.

Now she would convince her man that it's time they start trying for children. After a passionate love making, once a sperm invades an egg, the real biological transformation begins. During the conception both the male and female brains go through immense hormonal changes.

However, women are the leading figure in this journey. Two weeks after the fertilization, the fetus is implanted in the uterine lining. Thereafter the mother's blood supply gets attached with the fetus, which means now her body has to come up with blood supply for technically two bodies. This changes everything inside the female brain.

For starters, her progesterone level goes up. This is the reason why she feels sleepy all the time. High level of Progesterone acts as a sedative on a mother-to-be. Soon her breasts become tender and she needs to rest more than usual. And because she now needs to produce more blood than usual, her brain circuits for thirst and hunger start to run at full throttle. As a result she now always craves for food and water. Naturally she has to pee a lot now. She grows hankering for pickles and ice-cream.

During the first three months of pregnancy her brain circuits for sensitivity to smells get extremely active. This causes the disgusting sensation of nausea especially in the morning. By the fourth month, the female brain gets used to the massive transition of brain states. Her brain registers every single motion of her baby in her womb and slowly she becomes aware of the fact that she's growing a new life inside her belly. In the process her brain turns off the mechanism inside Hypothalamus responsible for menstruation.

There is another very perplexing thing that goes on in the female body during her late pregnancy. During this time, the placenta and fetus produce the stress hormones such as cortisol and adrenalin in large quantities. Then why on earth, such a huge quantity of stress hormones don't induce a feeling of stress in women during their late pregnancy. It seems pretty unlikely for cortisol and adrenalin.

Well, there is a fascinating biological reason behind it. First what you need to know is that the function of high levels of stress hormones during pregnancy is not to make a woman stressed, rather they make her vigilant about the safety of her baby. Then how does the brain counteract the innate stressful impact of the stress hormones! Here comes the amazing response of high levels of pregnancy hormones such as progesterone and estrogen. The sedative effect of these hormones creates a calm and peaceful reality for the expecting mother. As a result the biggest biological wonder of a woman's life becomes one of the most blissful experiences of womanhood.

And actually throughout pregnancy the male brain as well goes through typical paternal changes in parallel with the mother-to-be. Mother Nature programmed the male brain in such a way, that when his mate is pregnant, his aggressive testosterone level goes down and the driving force of paternal instincts, i.e. prolactin goes up. We believe that the brain circuits of paternal instincts are triggered by the motherly pheromones exuded from the sweat glands of the mother-to-be. In this context, some scientists believe that these hormonal changes in an expecting father, might be the cause of "couvade syndrome" or "sympathetic pregnancy". It's a widely documented syndrome, in which an expecting father experiences the same symptoms of pregnancy as his mate. The common symptoms experienced by many fathers-to-be (around 65% fathers worldwide) are increased appetite, weight gain, headache, nausea, breast augmentation, hardening of the nipples and insomnia etc. It's not at all a disease to be concerned about. Rather it's an evolutionary trick to transform a romantic lover into a caring father.

However, I believe there is a lot more going on behind sympathetic pregnancy than just the paternal hormones. It makes more sense if we include the impact of the

Mirror Neuron System (MNS). It is very much plausible, that due to the impact of high levels of prolactin as the male brain gets ready for the arrival of a new member into the family, the Mirror Neuron System gets highly active in order to sense every single emotional cue from the mate. Ergo, the MNS mirrors the motherly symptoms into the father-to-be and gives him the sense of carrying a new life. As a result, the bonding between the expecting father and mother grows even stronger. Expecting daddies who experience sympathetic pregnancy, actually have higher prolactin level than other daddies. Prolactin is the most crucial daddy hormone that prepares the soil of a paternal mind which is always cautious about the safety of the child and the mother.

Testosterone has the characteristic of suppressing both the maternal and paternal behaviors. That's the human biology which is designed in such a way that during pregnancy the motherly pheromones from the skin of an expecting mother not only increase the level of prolactin in her man, but also they reduce the testosterone level to a great extent.

Due to this lack of testosterone expecting daddies lose their sex-drive. Ergo, they don't feel comfortable to have sex during pregnancy. In women however, libido fluctuates during pregnancy. So at times they might find coitus more pleasurable or less pleasurable. A very common question that many expecting couples ask is *"Is it safe to have intercourse during pregnancy?"* And the simple answer is "Yes" (except in case of certain clinical conditions that may require abstinence from sex). And not only that. In fact, having intercourse during pregnancy until delivery time, is actually good for the baby. You must remember one thing that during pregnancy the baby's blood supply comes from the mother's body. So, when a couple have intercourse during pregnancy (taking their comfort into consideration), all the feel-good and bonding chemicals that are secreted into the blood stream of the mother, also marinate into the fetus. So, all the bliss that an expecting mommy feels, is naturally felt by her little one as well. And if an expecting father is worried that he might harm the baby, then what he needs to know is that he won't hurt the baby by making love. The baby is always protected by the amniotic sac and the strong muscles of the uterus. Added to that, the thick mucus plug that seals the cervix, safeguards against infection. When a baby receives love right inside the womb, it'll be born as a happy child and will always carry a cheerful vibe.

And finally the day comes, when a new life enters a couple's life. The water breaks and the amniotic fluid flows down the legs. The fetus itself has the innate ability to interact with the mother's body. A fully developed fetus sends signals to the mother's brain that it is ready to see the light of this world. Ergo, the labor starts. Due to the organized communication between the fetus and the mother's body, once the baby is ready to be born, suddenly the progesterone level drops and oxytocin floods the mother's biology. Oxytocin is really a versatile hormone. Not only it creates the bonding between mother and child, but also during labor it causes the uterus to start contracting.

Three of the major hormones that are involved with child-birth are oxytocin, beta-endorphin and adrenaline. These three work together harmoniously in regulating labor and birth. During childbirth the explosion of Oxytocin activates more receptor cells in the brain than ever before. This induces a euphoric and warm sensation in the mother while giving birth to her child. Oxytocin evokes powerful uterine contractions that allow the cervix to dilate in order to move the baby out of the birth canal and thereafter expel the placenta.

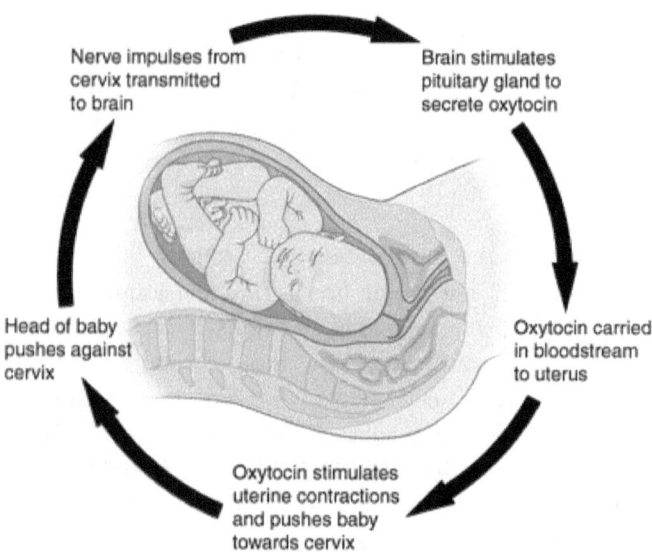

Figure 6.1 Positive Feedback Mechanism of Oxytocin Release

A fascinating biological process goes behind the release of Oxytocin during labor. Oxytocin is controlled by a positive feedback mechanism which means the release of the hormone causes an action which stimulates more of its own release. As the baby starts to come out through the birth canal its pressure against the cervix, and then against the tissues in the pelvic floor sends nerve impulses to the mother's brain to release more Oxytocin. In this way, further release of Oxytocin increases the contractions in intensity and frequency.

During late pregnancy often many women experience slight contractions before the actual onset of labor. These contractions are called "Braxton Hicks contractions", which are very common in the last week of pregnancy. Often first time expecting mothers mistake this "false labor" as the actual one. Then what is the way to tell if the contractions are real? With the onset of true labor contractions become regular, stronger, and more frequent. Whereas Braxton Hicks contractions do not settle in a regular pattern. Often, a change in activity, such as walking or lying down, makes Braxton Hicks contractions go away. However, it is always good to be sure by consulting your doctor.

The process of childbirth is usually divided into three stages of labor :

1. Dilation of the Cervix
2. Descent and Birth of the Baby
3. Afterbirth or Expulsion of the Placenta

The first stage of labor begins with regular contractions of the uterus. This lasts about 12 to 18 hours and ends when the cervix is fully dilated to 10 cm, so that the baby is ready to come out through the birth canal. In most cases, the baby's head enters the pelvis facing one side, and then rotates to face down. Sometimes, a baby will be facing up, towards the mother's abdomen. Based on the baby's position the doctor might rotate it. If a baby is not head down and its feet or buttocks are in position to come out first, then it is called "breech delivery".

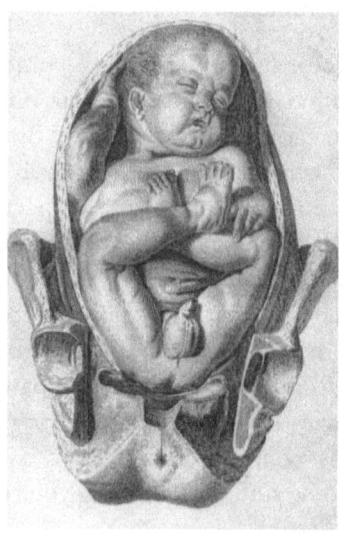

Figure 6.2 Baby's position during Breech Delivery (Illustration by William Smellie, 1792)

Studies have shown increased risks of morbidity, other health issues and mortality for breech delivery. That's why many hospital policies do not permit vaginal breech delivery. In this case, the doctor tries to rotate the baby, or suggests cesarean delivery. At the end of the first stage of labor, the contractions become longer and stronger.

The second stage is all about pushing and giving birth to the baby. It lasts around 20 minutes to 2 hours. The baby travels down the vagina and out into the world. For the first time mothers, this lasts up to about two hours. While giving birth, depending on the comfort level a woman can be in various positions such as squatting, sitting, kneeling and lying back. This is the hardest part for a mother-to-be. She needs to push really hard during contractions and relax between the contractions. It makes a woman really exhausted and she often feels unimaginable (by a man) nausea and trembling. It's an utterly strenuous experience for her. So it's very much necessary that her man is at all times right next to her.

However, Mother Nature has done everything in her capacity to ease the pain of child-birth. In response to the pain and stress, another fantastic hormone beta-endorphin is secreted into the blood stream. It is the body's natural pain-killer that has an amazing effect of relieving the mother-to-be from her utter stress and pain of pushing her child out of her. Now let's look into the fascinating craftsmanship of Mother Nature in preparing the brain circuits of a mother. She thought of every single scenario that might endanger the survival of the mother and her baby. If the pain of labor crosses the level of tolerance, it triggers more release of beta-endorphin that causes the oxytocin level to drop and therefore the contractions slow down slightly. In this way, the pain and stress of childbirth is naturally kept at a level that's bearable for the mother. Also, the fight-or-flight hormone, adrenalin adds some extra insurance to the survival of the mother and child. It keeps a woman alert and cautious while pushing.

When the top of the baby's head appears (crowning) through the vagina, the mother has to start pushing. If necessary, the doctor makes a small cut, called "episiotomy" to enlarge the vaginal opening, but it is very rare. Sometimes, the delivery is assisted with forceps or suction. After this tremendous ordeal, finally the new fellow arrives in this world. Then the umbilical cord is severed with the consent of the parents. However, some parents choose to omit umbilical cord severance entirely. This practice is called "lotus birth". The entire umbilical cord is left intact and allowed to dry and fall off on its own (typically on the 3rd day after birth). This kind of stone-age practice is highly dangerous as it increases the risk of infection in the baby.

Ultimately, it is oxytocin that triggers the "fetal ejection reflex", the final series of muscular contractions that push the baby out. The high level of oxytocin strengthens the mother-child relationship.

The cocktail of beta-endorphin, oxytocin and dopamine in the blood stream produces an altered state of consciousness that helps a woman flow with the process of child-birth even if it is long and strenuous. Despite all the extreme hardship of labor, the hormonal cocktail makes childbirth an extremely euphoric and blissful experience of a woman's life.

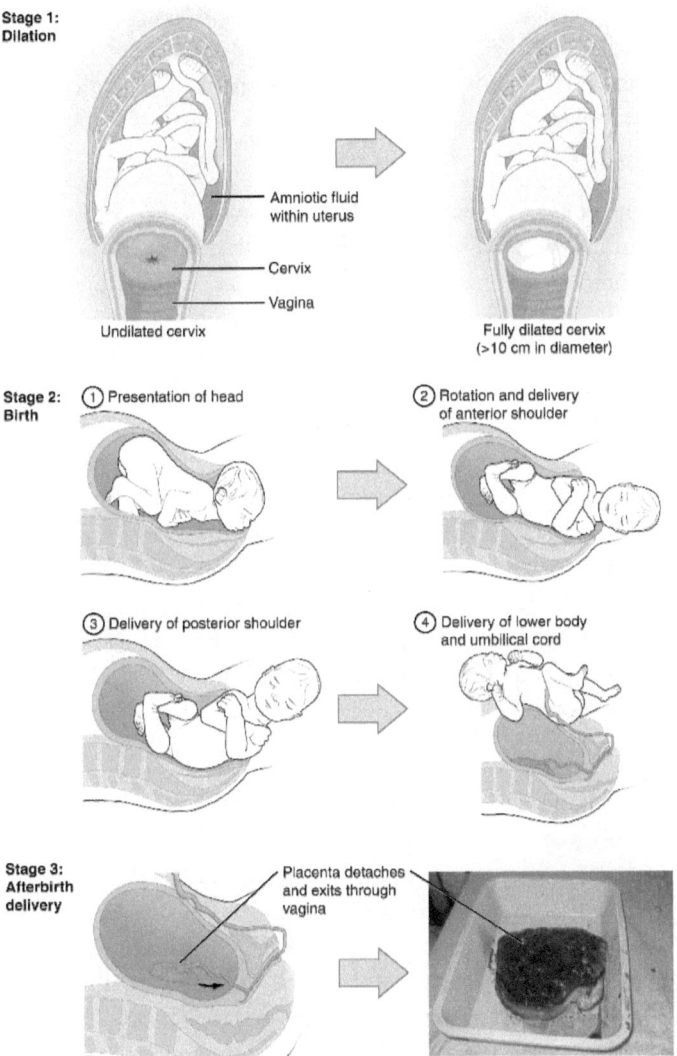

Figure 6.3 Stages of Child-birth

The third stage is called afterbirth. Contractions begin 5 to 30 minutes after birth, signaling that it's time to deliver the placenta. It often gives the mother chills and trembling. Once the placenta is delivered labor is officially over.

The extreme explosion of oxytocin and dopamine during the fetal ejection heightens the sense of hearing, touch, sight and smell in the mother. This allows the mother to register the unforgettable odor of her newborn instantly after birth. The lovely smell of her newborn's body becomes chemically embedded into the brain circuits of the new mother with utmost accuracy. Every single motion of her baby triggers fantabulous sensation in her body. Oxytocin enables the breasts to produce milk and makes the mother extremely sensitive to her baby's cry. Now her evolutionary instincts of motherhood run at full blast. Within hours after giving birth, the extremely protective motherly brain-biology takes hold of her very existence. Now there is only one evolutionary purpose of her existence, that is ensuring the survival of her infant.

Expecting daddies show significant increase of prolactin level weeks before birth. This is the hormone that rewires a lover male brain into a nurturing father brain. Also, the stress hormone cortisol doubles in level and makes a father more alert. Throughout pregnancy, the man's testosterone level starts to drop and in the first few weeks after birth it plummets by a third. On the contrary, the motherly hormone estrogen goes way up in the father than usual. These hormonal changes enable the father brain to bond with the new little offspring. In fact, they provide the new daddies some of the motherly instincts, such as being sensitive to the baby's cry. They may be not as good as the mommies but good enough to nurture the baby. And as I said earlier, lower testosterone level means decreased libido. Ergo, sex takes the back seat for a while. Now, the priority for both the mommy and daddy is to nurture the baby.

Physical contact is a great factor in the bonding process between the father and his newborn. Closeness increases the intensity of paternal love and builds new connections in the fatherly brain circuits. Physical interaction with the newborn actually increases the level of oxytocin in the father, which makes the bond even stronger. For a long time we thought that oxytocin plays a crucial role in motherhood. But we didn't have significant data about its impact on fatherhood until recently. A study by Gordon and colleagues (2010) on 80 couples who were first time parents, showed that at two points of time throughout parenthood after birth (6 weeks and 6 months postpartum) the levels of oxytocin in fathers are not different from the levels in their mates. So, while oxytocin floods in the mother's body due to childbirth and thereafter due to breastfeeding, in a father it is released more and more due to physical proximity with the baby.

However, the mother is the guide to this newly discovered parenthood in a relationship. She promotes all efforts of the father to get engaged with the new child as much as possible. As she is already full of oxytocin, she gets even closer to her man. And the bottom-line is that this is the perfect opportunity for all three to bond as a family in the most blissful manner. The parental brain circuits guide the parents throughout the whole process of parenthood.

* * *

CHAPTER 7

GUIDING BIOLOGY

Mother Nature programmed every single instinct inside our brain circuits with perfect accuracy. And each circuit turns on the other in perfect order, so that they serve a single purpose, i.e. survival of the species. Like a concerned mother, our Mother Nature has been rough on us during our earlier developmental stage as a species. Compelling us to go through the harsh primitive environment she made us more and more developed. And today we are the smartest species on planet earth and even in this entire solar system. I'm not sure about outside our solar system. But on this planet over the period of evolution we humans have learnt to tame the natural forces and even manipulate them at our best interest.

However when it comes to taming our instincts, the story is quite the opposite. Biological instincts are the driving force, or rather the guiding force of a species. Everything our ancestors learnt is encoded inside our instinctual blueprint.

Biological instincts are the key to understanding how every single human being is wired. The marvelous interplay of various brain circuits creates our instinctual

419

reality of the daily life. If you're conscious about the fact that there lies a complex yet vividly beautiful brain circuit mechanism behind every single impulse of your daily emotions, then you can choose how to react upon each of those impulses.

You can thus program your behavioral response in a certain situation. As long as we are humans, it is not in our capacity to fight with our biological instincts. In fact biology represents the very foundation of our personalities and behavioral responses. It is indeed the foundation of our very existence.

Once you understand and acknowledge the innate biological factors behind each behavior of yours and your partner's, you can choose whether or not to allow those behaviors to ruin your cheerful relationship. You can modulate your response based on the best interest of your relationship.

A fascinating thing about the human brain is its flexibility. Or in simple terms, it means nothing is completely fixed inside your brain. You are wired, but not hardwired. So, even though you are not able to completely alter your brain circuits, you still have the power to tame your behavioral impulses through scientific introspection. And in fact, upon achieving true awareness of your inner self, you'd get even closer to your partner.

Biology is not a compulsion, rather if understood properly, it can be utilized to kindle the fire even in a dying relationship. Thus, the brain biology hands over

to you the key to a healthy, compatible, lasting and
cheerful relationship.

* * *

BIBLIOGRAPHY

Book I
The Art of Neuroscience in Everything

Aserinsky E & Kleitman N. "Regularly occurring periods of eye motility, and concomitant phenomena, during sleep". Science 118, 1953.

Allman et. al. (2001). "The anterior cingulate cortex…" Ann N Y Acad Sci. 935

Apps M., Balsters J, Ramnani N, (2009). "Anterior Cingulate Cortex: Monitoring The Outcomes Of Others' Decisions", Royal Holloway University of London, London, United Kingdom, http://www.medicalnewstoday.com/articles/153472.php

Ardila A, Gomez J. Paroxysmal "feeling of somebody being nearby". Epilepsia 1988;

Babayev ES, Allahverdiyeva AA. Effects of geomagnetic activity variations on the physiological and psychological state of functionally healthy humans: some results of Azerbaijani studies. Adv Space Res 2007.

Burchett, Scott A. and Phillip T. Hicks. 2006. "The mysterious trace amines: Protean neuromodulators of synaptic transmission in mammalian brain." Progress in Neurobiology.

Balleine BW, Delgado MR, Hikosaka O (2007a). "The role of the dorsal striatum in reward and decision-making". J Neurosci 27

Barrett, D. (2001) The Committee of Sleep: How Artists, Scientists, and Athletes Use their Dreams for Creative Problem Solving—and How You Can Too. NY: Crown Books/Random House/hardback

Barrett, D. (2007) "An Evolutionary Theory of Dreams and Problem-Solving" in Barrett, D.L. & McNamara, P. (Eds.) The New Science of Dreaming, Volume III: Cultural and Theoretical Perspectives on Dreaming, NY, NY: Praeger/Greenwood, 2007.

Bechara, A. et al (1994) "Insensitivity to future consequences following damage to human prefrontal cortex". Cognition 50

Blackmore, Susan (2004). Consciousness an introduction. New York, NY: Oxford University Press

Braun, AR (1997) "Regional cerebral blood flow throughout the sleep-wake cycle. An H2(15)O PET study". Oxford Journals, Brain; (1997) 120 (7).doi: 10.1093/brain/120.7.1173

Bottini G, Corcoran R, Sterzi R, Paulesu E, Schenone P, Scarpa P, et al. (1994) "The role of the right hemisphere in the interpretation of figurative aspects of language". Brain 117

Botvinick M, et al (1999) "Conflict monitoring versus selection-for-action in ACC" Nature 402 (6758): 179–81

Bush G, Luu P, Posner MI. (2000). "Cognitive and emotional influences in anterior cingulate cortex". Trends Cogn Sci. 2000 Jun; 4(6)

Blanke, O. and Arzy, S. "The Out-of-Body Experience: Disturbed Self-Processing at the Temporo-Parietal Junction" THE NEUROSCIENTIST 2005.

Blanke O, Ortigue S, Landis T, Seeck M. 2002. Stimulating illusory own-body perceptions. Nature.

Bonda E, Petrides M, Frey S, Evans A. 1995. Neural correlates of men- tal transformations of the body-in-space. Proc Natl Acad Sci U S A.

Brandt T. 2000. Central vestibular disorders. In: Vertigo: its multisensory syndromes. London: Springer.

Brugger P, Agosti R, Regard M, Wieser HG, Landis T. 1994. Heautoscopy, epilepsy, and suicide. J Neurol Neurosurg Psychiatr.

Brugger P, Regard M, Landis T. 1997. Illusory reduplication of one's own body: phenomenology and classification of autoscopic phenomena. Cogn Neuropsychiatr.

Breur, J. and Freud, S. (1893-1895) Studies on Hysteria.

Carta, M.G. "Women And Hysteria In The History Of Mental Health" Clinical Practice & Epidemiology in Mental Health, 2012

Cartwright, R. (1993). "Functions of Dreams". Encyclopedia of Sleep and Dreaming.

Calvert GA, Campbell R, Brammer MJ. 2000. Evidence from functional magnetic resonance imaging of crossmodal binding in the human heteromodal cortex. Curr Biol.

Campbell A. The limbic system and emotion in relation to acupuncture. Acupuncture in Medicine.1999.

Cook, R. "Mirror neurons: From origin to function" Behavioral and Brain Sciences 2014.

Daly DD. 1958. Ictal affect. Am J Psychiatry.

Darwin, Charles. "On the origin of species by means of natural selection" (original edition, 1859).

Darwin, Charles. "The Descent of Man" (original edition, 1871).

Decety J, Sommerville JA. 2003. Shared representations between self and other: a social cognitive neuroscience view. Trends Cogn Sci.

Denning TR, Berrios GE. 1994. Autoscopic phenomena. Br J Psychiatry.

Devinsky O, Feldmann E, Burrowes K, Bromfield E. 1989. Autoscopic phenomena with seizures. Arch Neurol.

Downing PE, Jiang Y, Shuman M, Kanwisher N. 2001. A cortical area selective for visual processing of the human body. Science.

Dang-Vu et.al., (2007) "Neuroimaging of REM Sleep and Dreaming", In D. Barret & P. McNamara (Eds.),

The New Science of Dreaming: Vol. 1. Bilogical Aspects. Westport Connecticut, Praeger.

Dastur H.M., Desai A.D., "A comparative study of brain tuberculomas and gliomas based upon 107 case records of each". Brain. 1965

Devinsky O, Lai G. Spirituality and religion in epilepsy. Epilepsy Behav 2008.

Dostoyevsky, The Possessed (original edition 1872)

Dostoyevsky, The Brothers Karamazov, (original edition 1880)

Dostoyevsky, The Insulted and Injured, (original edition 1861)

Dostoyevsky, The Idiot,. (original edition 1869)

Dotta BT, Persinger MA. "Doubling" of local photon emissions when two simultaneous, spatially separated, chemiluminescent reactions share the same magnetic field configurations. J Biophys Chem 2012

Dotta BT, Bucner CA, Lafrenie RM, Persinger MA. Photon emissions from human brain and cell culture exposed to distally rotating magnetic fields shared by separate light-stimulated brains and cells. Brain Res 2011

Esquirol, Étienne (1838). Baillière, Jean-Baptiste (and sons), ed. Des maladies mentales considérées sous les rapports médical, hygiénique et médico-légal, [Mental illness as considered in medical, hygienic, and medico-legal reports] Volume 1 and 2.

Esquirol, Étienne 1845. Mental maladies; a treatise on insanity (original French edition 1838).

Esch, T. and Stefano, G.B "The Neurobiology of Love" Neuroendocrinology Letters No.3 June Vol.26, 2005.

Esch T. [Health in stress: Change in the stress concept and its significance for prevention, health and life style]. Gesundheitswesen 2002.

Flaubert Gustave, Madame Bovary 1856

Feinstein, D. (1990) "The Dream as a Window to Your Evolving Mythology," in S. Krippner, Dreamtime & Dreamwork, Jeremy P. Tarcher Inc. Los Angeles, CA

Farrell MJ, Robertson IH. 2000. The automatic updating of egocentric spatial relationships and its impairment due to right posterior cortical lesions. Neuropsychologia.

Fasold O, von Bevern M, Kuhberg M, Ploner CJ, Vilringer A, Lempert T, Wenzel R. 2002. Human vestibular cortex as identified with caloric vestibular stimulation by functional magnetic resonance imaging. Neuroimage.

Fiss, (1986)., "An experimental self-psychology of dreaming" Journal Of Mind And Behavior 1986

Foulkes, D. (1982) Children's dreams: longitudinal studies, Wiley, 1982

Fosse MJ, Fosse R, Hobson JA, Stickgold RJ. (2003). "Dreaming and episodic memory: a functional dissociation?" J Cogn Neurosci. 2003 Jan 1

Freud, S. (1900). The Interpretation of Dreams

Freud, S. "Selected papers on hysteria and other psychoneuroses" Journal of Nervous and Mental Disease 1909.

Freud, S. "The Origin and Development of Psychoanalysis" (1910)

Freud, S. (1914) Psychopathology of everyday life

Goodwin GM, McCloskey DI, Matthews PBC. 1972. Proprioceptive illusions induced by muscle vibration: contribution by muscle spindles to perception? Science.

Green CE. 1968. Out-of-body experiences. London: Hamish Hamilton.

Grossman E, Donnelly M, Price R, Pickens D, Morgan V, Neighbor G, and others. 2000. Brain areas involved in perception of biological motion. J Cogn Neurosci.

Grüsser OJ, Landis T. 1991. The splitting of "I" and "me": heautoscopy and related phenomena. In: Visual agnosias and other disturbances of visual perception and cognition. Amsterdam: MacMillan.

Greenberg, R., Pearlman, C. (1975). "Rem Sleep and the Analytic Process", Psychoanal Q., 44

Greene, G. (2010) "The Power And Purpose Of Dreams", Psychology Today Published on February 15, 2010

Griffin, J. (1997). "The Origin of Dreams: How and why we evolved to dream". The Therapist 4

Gribbin John. (2013) "Erwin Schrodinger and The Quantum Revolution", Wiley

Goetz CG (August–September 2009). "Jean-Martin Charcot and movement disorders: neurological legacies to the 21st century". International Parkinson and Movement Disorder Society. Retrieved 2013.

Gusnard D., Akbudak E., Shulman G., Raichle M. (2001). Proceedings of the National Academy of Science, March 27, 2001 vol. 98 no. 7

Greyson, Bruce. 2006. "Near-Death Experiences and Spirituality." Zygon: Journal of Religion and Science.

Geschwind N. "Behavioural changes in temporal lobe epilepsy". Psychol Med. 1979.

Hatfield E. Love, Sex and Intimacy. New York: Harper Collins 1993.

Harribance CC. Sean Harribance: A psychic predicts the future. Port of Spain: Sean Harribance Institute; 1994.

Heaton JP, Adams MA. Update on central function relevant to sex: remodeling the basis of drug treatments for sex and the brain. Int J Impot Res 2003.

Hartmann, E. (1995). "Making connections in a safe place: Is dreaming psychotherapy?" Dreaming 5

Hartmann, E. (2011). The Nature and Functions of Dreaming, Oxford University Press; also Hartmann, E. (2012). "On the nature and functions of dreaming"; http://www.ceoniric.cl/english/articles/on_the_nature _and_functions.htm

Hayden, B., Pearson, J., & Platt, M. (2009). "Fictive reward signals in the anterior cingulate cortex". Science, 324, 948–950. doi:10.1126/science.1168488

Hobson, J. A., Pace-Schott, E. F., Stickbold, R. (2003). "Dreaming and the brain: toward a cognitive neuroscience of conscious states". In E.F. Pace-Schott, M. Solms, M. Blagrove, S. Harnad (Eds.), Sleep and Dreaming. New York, USA, Cambridge University Press

Hobson, J.A.; McCarley, R. 1977. "The brain as a dream state generator: an activation-synthesis hypothesis of the dream process". American Journal of Psychiatry, 134

Hobson, J. A. (2009). Journal Nature Reviews Neuroscience, Oct 2009

Hobson, J.A. (2009). "REM sleep and dreaming: towards a theory of protoconsciousness". Nature Reviews 10 (11): 803–813. doi:10.1038/nrn2716. PMID 19794431

Horton, C., Christopher J. A., et.al. (2009). Consciousness and Cognition 18 (3)

Hoss, R. (2005). Dream Language: Self-Understanding through Imagery and Color. Ashland: Innersource

Halligan PW. 2002. Phantom limbs: the body in mind. Cogn Neuropsychiatr.

Halligan PW, Fink GR, Marshal JC, Vallar G. 2003. Spatial cognition: evidence from visual neglect. Trends Cogn Sci.

Harribance" International Journal of Yoga Vol. 5, 2012.

Harrington A. Unfinished business: models of laterality in the nineteenth century. In: Davidson RJ, Hugdhal K, editors. Brain asymmetry. MIT Press; 1995.

Hoss Robert J. "The Neuropsychology of Dreaming: Studies and Observations" 2013.

Hécaen H, Ajuriaguerra J. 1952. L'Héautoscopie. In: Méconnassiances et hallucinations corporelles. Paris: Masson.

Hécaen H, Green A. 1957. Sur l'héautoscopie. Encephale.

Irwin HJ. 1985. Flight of mind: a psychological study of the out-of- body experience. Metuchen (NJ): Scarecrow Press.

Jung (1945). "On the Nature of Dreams" (1945). In Collected Works 8: The Structure and Dynamics of the Psyche.

Jung, C. (1971), The Portable Jung, J. Campbell (Ed.), New York: Viking Press.

Jung, C. G. (1973). Man and His Symbols, New York, Dell Publishing.

Krippner S, Persinger M. Evidence for enhanced congruence between dreams and distant target material during periods of decreased geomagnetic activity. J Sci Explor 1996

Komisaruk BR, Whipple B. Love as sensory stimulation: physiological consequences of its deprivation and expression. Psychoneuroendocrinology 1998

Kölmel HW. 1985. Complex visual hallucinations in the hemianopic field. J Neurol Neurosurg Psychiatry.

Lackner JR. 1988. Some proprioceptive influences on the perceptual representation of body shape and orientation. Brain.

Litovitz TA, Penafiel M,Krause D,Zhang D,Mullins JM.The roleoftemporal sensing in bioelectromagnetic effects. Bioelectromagnetics 1997.

Lagace N, St-Pierre LS, Persinger MA. Attenuation of epilepsy-induced brain dam- age in the temporal cortices of rats by exposure to LTP-patterned magnetic fields. Neurosci Lett 2009.

Lacoboni, M. and Dapretto, M. "The mirror neuron system and the consequences of its dysfunction" Nature 2005.

Lackner JR. 1992. Sense of body position in parabolic flight. Ann N Y Acad Sci.

Leube DT, Knoblich G, Erb M, Grodd W, Bartels M, Kircher TT. 2003. The neural correlates of perceiving one's own movements. Neuroimage.

Laidler Keith J. 2002, "Energy and The Unexpected", Oxford University Press

Lippman CW. 1953. Hallucination of physical duality in migraine. J Nerv Ment Dis.

Lobel E, Kleine J, Leroy-Wilig A. 1999. Functional MRI of galvanic vestibular stimulation. J Neurophysiol.

Lunn V. 1970. Autoscopic phenomena. Acta Psychiatr Scand 46(Suppl 219).

Mulligan BP, Hunter MD, Persinger MA. Effects of geomagnetic activity and atmospheric power variations on quantitative measures of brain activity: replication of the Azerbaijani studies. Adv Space Res 2010.

Moody, Paul. 1975. Life after Life: The Investigation of a Phenomenon—Survival of Bodily Death . Atlanta: Mockingbird Books.

Martens, P.R. 1994. "Near-Death Experiences in Out-of-Hospital Cardiac Arrest Survivors. Meaningful Pheneomena or just Fantasy of Death?" Resuscitation.

Melzack R. 1990. Phantom limbs and the concept of a neuromatrix. Trends Neurosci.

Metzinger T. 2003. Being no one. Cambridge (MA): MIT Press.

Mittelstaedt H, Glasauer S. 1993. Illusions of verticality in weightlessness. Clin Invest.

Neisser U. 1988. The five kinds of self-knowledge. Phil Psychol.

Newberg, A. "Cerebral blood flow changes associated with different meditation practices and perceived depth of meditation" Psychiatry Research: Neuroimaging 2010.

Onions CT. The Oxford Dictionary of English Etymology. New York: Oxford University Press 1966.

Persinger, "'I would kill in God's name' role of sex, weekly church attendance, report of a religious experience and limbic lability" Perceptual and Motor Skills 1997.

Persinger "Experimental simulation of the God experience" Neurotheology 2003.

Persinger, Corradini, Clement, Keaney, et al "Neurotheology and its convergence with neuroquantology" NeuroQuantology 2010.

Persinger, Koren and St-Pierre "The electromagnetic induction of mystical and altered states within the laboratory" Journal of Consciousness Exploration and Research 2010.

Persinger "Case report: A prototypical spontaneous 'sensed presence' of a sentient being and concomitant electroencephalographic activity in the clinical laboratory" Neurocase 2008.

Persinger and Saroka "Potential production of Hughlings Jackson's "parasitic consciousness" by physiologically-patterned weak transcerebral magnetic fields: QEEG and source localization" Epilepsy & Behavior 28 (2013).

Persinger. "The neuropsychiatry of paranormal experiences". J Neuropsychiatry Clin Neurosci 2001.

Persinger, M. "Billions of Human Brains Immersed Within a Shared Geomagnetic Field: Quantitative Solutions and Implications for Future Adaptations" The Open Biology Journal, 2013.

Persinger MA, Lavallee CF. Theoretical and experimental evidence of macroscopic entanglement between human brain activity and photon emissions; implications for quantum consciousness and future applications. J Cons Explor Res 2010

Persinger MA, Lavallee CF. The sum of N=N and the quantitative support for the cerebral holographic and electromagnetic configuration of consciousness. J Cons Stud 2012

Persinger MA. 10-20 Joules as a neuromolecular quantum in medicinal chemistry: an alternative to the myriad molecular pathways? Curr Med Chem 2010

Persinger MA. On the possible representation of the electromagnetic equivalents of all human memory within the earth's magnetic field: implications for theoretical biology. Theor Biol Insights 2008

Persinger, MA. "Schumann Resonance Frequencies Found Within Quantitative Electroencephalographic Activity: Implications for Earth-Brain Interactions" International Letters of Chemistry, Physics and Astronomy 2014.

Persinger, MA. "Quantitative Evidence for Direct Effects Between Earth-Ionosphere Schumann Resonances and Human Cerebral Cortical Activity"

International Letters of Chemistry, Physics and Astronomy 2014.

Persinger, MA. "Terrestrial and lunar gravitational forces upon the mass of a cell: relevance to cell function" International Letters of Chemistry, Physics and Astronomy 2014.

Persinger, MA. "Dream ESP Experiments and Geomagnetic Activity" Journal of American Society for Psychical Research Vol 83, 1989.

Persinger, MA. and Saroka, KS. "Protracted parahippocampal activity associated with Sean Persinger MA, Krippner S. Dream ESP experiments and geomagnetic activity. J Am Soc Psychical Res 1989

Persinger "Experimental Facilitation of the Sensed Presence: Possible Intercalation between the Hemispheres Induced by Complex Magnetic Fields" Journal of Nervous and Mental Disease 2002.

Palmer J. 1978. The out-of-body experience: a psychological theory. Parapsychol Rev.

Persinger MA. Geophysical variables and behaviour: LXXI. Differential contribution of geomagnetic activity to paranormal experiences concerning death and crisis: An alternative to the ESP hypothesis. Percept Motor Skills 1993

Persinger MA, Saroka KS, Lavallee CF, Booth JN, Hunter MD, Mulligan BP, et al. Correlated cerebral events between physically and sensory isolated pairs of subjects exposed to yoked circumcerebral magnetic fields. Neuroscience Lett 2010

Persinger MA, Roll WG, Tiller SG, Koren SA, Cook CM. Remote viewing with the artist Ingo Swann: Neuropsychological profile, electroencephalographic correlates, magnetic resonance imaging (MRI) and possible mechanisms. Percept Motor Skills 2002

Plato, Phaedrus 370 BC

Revonsuo, A. (2000). "The reinterpretation of dreams: an evolutionary hypothesis of the function of dreaming". Behavioral Brain Science 23.

Ryan A. New insights into the links between ESP and geomagnetic activity. J Sci Explor, 2008

Roth, G and Dicke, U,(2005) Evolution of the Brain and the Intelligence, TRENDS in Cognitive Sciences.

Ruby P, Decety J. 2001. Effect of subjective perspective taking during simulation of action: a PET investigation of agency. Nat Neurosci.

Rizzolatti, G. & Fadiga, L. (1998) Grasping objects and grasping action meanings: The dual role of monkey rostroventral premotor cortex (area F5). Novartis Foundation Symposium 218

Rizzolatti, G., Fadiga, L., Gallese, V. & Fogassi, L. (1996) Premotor cortex and the recognition of motor actions. Social Cognitive and Affective Neuroscience 3

Rizzolatti, G., Fogassi, L. & Gallese, V. (2001) Neurophysiological mechanisms underlying the understanding and imitation of action. Nature Reviews Neuroscience 2

Rizzolatti G., Fogassi L. & Gallese V. (2004) Cortical mechanism subserving object grasping, action understanding and imitation. In: The cognitive neurosciences, 3rd edition, ed. M. S. Gazzaniga, A Bradford Book/MIT Press

Rizzolatti, G. & Arbib, M. A. (1998) Language within our grasp. Trends in Neurosciences 21 [aRC, LLH]

Rizzolatti, G., Camarda, R., Fogassi, L., Gentilucci, M., Luppino, G. & Matelli, M. (1988) Functional organization of inferior area 6 in the macaque monkey. II. Area F5 and the control of distal movements. Experimental Brain Research 71

Rizzolatti, G. & Luppino, G. (2001) The cortical motor system. Neuron 31 [LF] Rizzolatti, G. & Matelli, M. (2003) Two different streams form the dorsal visual system: Anatomy and functions. Experimental Brain Research 153

Rizzolatti, G. & Sinigaglia, C. (2008) Mirrors in the brain. How our minds share actions and emotions. Oxford University Press.

Rizzolatti, G. & Sinigaglia, C. (2010) The functional role of the parieto-frontal mirror circuit: Interpretations and misinterpretations. Nature Reviews Neuroscience 11

Ratnasuriya, R.H. "Joan of Arc, creative psychopath: is there another explanation?" Journal of The Royal Society of Medicine 1986.

Ramachandran, V. S. (2000) Mirror neurons and imitation learning as the driving force behind "the great leap forward" in human evolution. Edge69. [Available Online at: http://www.edge.org/3rd_culture/ramachandran/rama chandran_in- dex.html]

Ramachandran,V. S. (2009) The neurons that shaped civilization. Available at: http:// www.ted.com/talks/vs_ramachandran_the_neurons_th at_shaped_civilization.Html

Rocca, M. A., Tortorella, P., Ceccarelli, A., Falini, A., Tango, D., Scotti, G., Comi, G. & Fillipi, M. (2008) The "mirror-neuron system" in MS: A 3 tesla fMRI study. Neurology 70

Rochat, M. J., Caruana, F., Jezzini, A., Escola, L., Intskirveli, I., Grammont, F., Gallese, V., Rizzolatti, G. & Umiltà, M. A. (2010) Responses of mirror neurons in area F5 to hand and tool grasping observation. Experimental Brain Research 204

Rochat, M. J., Serra, E., Fadiga, L. & Gallese, V. (2008) The evolution of social cognition: Goal familiarity shapes monkeys' action understanding. Current Biology 18

Rochat, P. (1998) Self-perception and action in infancy. Experimental Brain Research 123

Rosenbaum, D. (1991) Human motor control. Academic Press.

Roth, T. L. (2012) Epigenetics of neurobiology and behavior during development and adulthood. Developmental Psychobiology 54. doi: 10.1002/dev.20550.

Rushworth, M. F., Mars, R. B. & Sallet, J. (2013) Are there specialized circuits for social cognition and are they unique to humans? Current Opinion in Neurobiology 23

Russell, J. L., Lyn, H., Schaeffer, J. A. & Hopkins, W. D. (2011) The role of socio- communicative rearing environments in the development of social and physical cognition in apes. Developmental Science 14

Sampson, G. (2002) Exploring the richness of the stimulus. The Linguistic Review 19

Sanefuji, W. & Ohgami, H. (2013) "Being-imitated" strategy at home-based inter- vention for young children with autism. Infant Mental Health Journal 34. doi: 10.1002/imhj.21375.

Santos, L. R., Nissen, A. G. & Ferrugia, J. A. (2006) Rhesus monkeys, Macaca mulatta, know what others can and cannot hear. Animal Behaviour 71

Singer, T. et. al. "Empathy for Pain Involves the Affective but not Sensory Components of Pain" Science 303 (2004).

Sheils D. 1978. A cross-cultural study of beliefs in out-of-the-body experiences, waking and sleeping. J Soc Psych Res.

Strassman, R. "DMT: The Spirit Molecule" 2001.

Smith BH. 1960. Vestibular disturbances in epilepsy. Neurol.

Schrodinger Erwin. (2012) "What is Life?: With Mind and Matter and Autobiographical Sketches", Cambridge University Press

Schumann W. O. (1952). "Über die strahlungslosen Eigenschwingungen einer leitenden Kugel, die von einer Luftschicht und einer Ionosphärenhülle umgeben ist". Zeitschrift und Naturfirschung 7a. Bibcode:1952ZNatA...7..149S. doi:10.1515/zna-1952-0202.

Schumann W. O. (1952). "Über die Dämpfung der elektromagnetischen Eigenschwingnugen des Systems Erde – Luft – Ionosphäre". Zeitschrift und Naturfirschung 7a. Bibcode:1952ZNatA...7..250S. doi:10.1515/zna-1952-3-404.

Schumann W. O. (1952). "Über die Ausbreitung sehr Langer elektriseher Wellen um die Signale des Blitzes". Nuovo Cimento 9. doi:10.1007/BF02782924.

Schumann W. O. & H. König (1954). "Über die Beobactung von Atmospherics bei geringsten Frequenzen". Naturwiss 41. Bibcode:1954NW.41.183S. doi:10.1007/BF00638174.

Shipley JT. Dictionary of Word Origins. New York: Philosophical Library 1945.

Stefano GB, Scharrer B. Endogenous morphine and related opiates, a new class of chemical messengers. Adv Neuroimmunol 1994

Stefano GB, Scharrer B, Smith EM, Hughes TK, Magazine HI, Bil- finger TV et al. Opioid and opiate immunoregulatory processes. Crit Rev in Immunol 1996.

Simpson JA, Rholes WS. Stress and secure base relationships in adulthood. In: Bartholomew K, Perlman D, editors. Advances in personal relationships (Vol. 5): Attachment processes in adult- hood. London: Kingsley 1994

Slingsby BT, Stefano GB. Placebo: Harnessing the power within. Modern Aspects of Immunobiology 2000

Slingsby BT, Stefano GB. The active ingredients in the sugar pill: Trust and belief. Placebo 2001

Small DM, Jones-Gotman M, Dagher A. Feeding-induced dopamine release in dorsal striatum correlates with meal pleasantness ratings in healthy human volunteers. Neuroimage 2003

Small DM, Zatorre RJ, Dagher A, Evans AC, Jones-Gotman M. Changes in brain activity related to eating chocolate: From pleasure to aversion. Brain 2001

Smith CM. Elements of Molecular Neurobiology. 3rd ed. New York: Wiley-Liss 2002

Sonetti D, Peruzzi E, Stefano GB. Endogenous morphine and ACTH association in neural tissues. Medical Science Monitor 2005

Spector S, Munjal I, Schmidt DE. Endogenous morphine and codeine. Possible role as endogenous anticonvulsants. Brain Res 2001

Spencer H. Principles of Psychology. New York: Appleton 1800

Stefano GB. Endocannabinoid immune and vascular signaling. Acta Pharmacologica Sinica 2000

Stefano GB, Benson H, Fricchione GL, Esch T. The Stress Re- sponse: Always good and when it is bad. New York: Medical Science International 2005

Stefano GB, Cadet P, Zhu W, Rialas CM, Mantione K, Benz D et al. The blueprint for stress can be found in invertebrates. Neuroendocrinology Letters 2002.

Sternberg, Robert J. (2007). "Triangulating Love". In Oord, T. J. The Altruism Reader: Selections from Writings on Love, Religion, and Science. West Conshohocken, PA: Templeton Foundation. ISBN 9781599471273.

Sternberg, Robert J. (2004). "A Triangular Theory of Love". In Reis, H. T.; Rusbult, C. E. Close Relationships. New York: Psychology Press. ISBN 0863775950.

Sternberg, Robert J. (1997). "Construct validation of a triangular love scale". European Journal of Social Psychology 27

Slater E, Beard AW. The schizophrenia-like psychoses of epilepsy. Br J Psychiatry 1963.

Tolstoy Leo, Anna Karenina 1877

Tan SY, Shigaki D (May 2007). "Jean-Martin Charcot (1825–1893): pathologist who shaped modern neurology". Singapore Med J 48. PMID 17453093

Tommasi MCO. Orgiasmo orgies and ritual in the ancient world: a few notes. Kervan 2006-2007

Turner, J. H. (2000b). On the origins of human emotions: A sociological inquiry into the evolution of human affect. Stanford, California: Stanford University Press.

Venkatasubramanian G, Jayakumar PN, Nagendra HR, Nagaraja D, Deeptha R, Gangadhar BN. Investigating paranormal phenomena: Functional brain imaging of telepathy. Int J Yoga 2008

Whinnery, J.E. 1997. "Psychophysiologic Correlates of Unconsciousness and near- death experiences." Journal of Near-Death Studies.

Watkins, K. & Paus, T. Modulation of motor excitability during speech perception: the role of Broca's area. J. Cogn. Neurosci.

Zacks JM, Ollinger JM, Sheridan MA, Tversky B. 2002. A parametric study of mental spatial transformations of bodies. Neuroimage.

BOOK II
YOUR OWN NEURON
A TOUR OF PSYCHIC BRAIN

Blanke, O. and Arzy, S. "The Out-of-Body Experience: Disturbed Self-Processing at the Temporo-Parietal Junction" THE NEUROSCIENTIST 2005.

Blanke O, Ortigue S, Landis T, Seeck M. 2002. Stimulating illusory own-body perceptions. Nature.

Bonda E, Petrides M, Frey S, Evans A. 1995. Neural correlates of mental transformations of the body-in-space. Proc Natl Acad Sci U S A.

Bomsel-Helmreich O, Al Mufi W. The phenomenon of monozygosity: spontaneous zygotic Spltting. In: Blickstein I. Multiple Pregnancy. Epidemiology, Gestation, and Perinatal Outcome. London: Taylor & Francis; 2005

Bierman DJ, Radin DI. Anomalous anticipatory response on randomized future conditions. Percept Mot Skills. 1997

Burchett, Scott A. and Phillip T. Hicks. 2006. "The mysterious trace amines: Protean neuromodulators of synaptic transmission in mammalian brain." Progress in Neurobiology.

Bierman DJ, Scholte HS. Anomalous anticipatory brain activation preceding exposure of emotional and neutral pictures. In: Proceedings of presented Papers: The,

Parapsychological Association 45th Annual Convention. Cary, NC: Parapsychological Association; 2002

Bem DJ. Feeling the future: experimental evidence for anomalous retroactive influences on cognition and affect. J Pers Soc Psychol. 2011

Boucsein W. Electrodermal Activity, Plenum Series in Behavioral Psychophysiology and Medicine. New York: Plenum Press; 1992

Blackmoore, Susan, J. and Chamberlain, F. "ESP and thought concordance in twins: A method of comparison", Journal of the Society of Psychical Research 1993

Brandt T. 2000. Central vestibular disorders. In: Vertigo: its multisensory syndromes. London: Springer.

Brugger P, Agosti R, Regard M, Wieser HG, Landis T. 1994. Heautoscopy, epilepsy, and suicide. J Neurol Neurosurg Psychiatr.

Brugger P, Regard M, Landis T. 1997. Illusory reduplication of one's own body: phenomenology and classification of autoscopic phe- nomena. Cogn Neuropsychiatr.

Calvert GA, Campbell R, Brammer MJ. 2000. Evidence from functional magnetic resonance imaging of crossmodal binding in the human heteromodal cortex. Curr Biol.

Charlesworth, E.A. "Psi and imaginary dreams" RIP 1974

Duane, T.D. and Behrendt, T. "Extrasensory Electroencephalographic Induction between Identical Twins" Science 1975

Dotta BT, Bucner CA, Lafrenie RM, Persinger MA. Photon emissions from human brain and cell culture exposed to distally rotating magnetic fields shared by separate light-stimulated brains and cells. Brain Res 2011.

Dotta BT, Persinger MA. "Doubling" of local photon emissions when two simultaneous, spatially separated, chemiluminescent reactions share the same magnetic field configurations. J Biophys Chem 2012.

Daly DD. 1958. Ictal affect. Am J Psychiatry.

Decety J, Sommerville JA. 2003. Shared representations between self and other: a social cognitive neuroscience view. Trends Cogn Sci.

Denning TR, Berrios GE. 1994. Autoscopic phenomena. Br J Psychiatry.

Devinsky O, Feldmann E, Burrowes K, Bromfield E. 1989. Autoscopic phenomena with seizures. Arch Neurol.

Downing PE, Jiang Y, Shuman M, Kanwisher N. 2001. A cortical area selective for visual processing of the human body. Science.

Esser AH, Etter TL, Chamberlain WB. Preliminary report: physiological concomitants of "communication" between isolated subjects. Int J Parapsychol. 1967

Farrell MJ, Robertson IH. 2000. The automatic updating of egocentric spatial relationships and its impairment due to right posterior cortical lesions. Neuropsychologia.

Fasold O, von Bevern M, Kuhberg M, Ploner CJ, Vilringer A, Lempert T, Wenzel R. 2002. Human vestibular cortex as identified with caloric vestibular stimulation by functional magnetic resonance imaging. Neuroimage.

Goodwin GM, McCloskey DI, Matthews PBC. 1972. Proprioceptive illusions induced by muscle vibration: contribution by muscle spindles to perception? Science.

Glicksohn J. Belief in the paranormal and subjective paranormal experience. Pers Individ Dif. 1990

Grinberg-Zylberbaum J, Delaflor M, Sanchez-Arellano ME, Guevara MA, Perez M. Human communication and the electrophysiological activity of the brain. Subtle Energies. 1993

Grinberg-Zylberbaum J, Delaflor M, Attie L, Goswami A. The Einstein-Podolsky-Rosen paradox in the brain: the transferred potential. Phys Essays. 1994

Greyson, Bruce. 2006. "Near-Death Experiences and Spirituality." Zygon: Journal of Religion and Science.

Gurney, E. "Phantasms of the Living" Volume 1 and 2, 1886

Gallup GH, Newport F. Belief in paranormal phenomena among adult Americans. Skeptical Inquirer. 1991

Green CE. 1968. Out-of-body experiences. London: Hamish Hamilton.

Grossman E, Donnelly M, Price R, Pickens D, Morgan V, Neighbor G, and others. 2000. Brain areas involved in perception of biological motion. J Cogn Neurosci.

Grüsser OJ, Landis T. 1991. The splitting of "I" and "me": heautoscopy and related phenomena. In: Visual agnosias and other disturbances of visual perception and cognition. Amsterdam: MacMillan.

Halligan PW. 2002. Phantom limbs: the body in mind. Cogn Neuropsychiatr.

Halligan PW, Fink GR, Marshal JC, Vallar G. 2003. Spatial cognition: evidence from visual neglect. Trends Cogn Sci.

Haraldsson E. "Representative national surveys of psychic phenomena: Iceland, Great Britain, Sweden, USA and Gallup's multinational survey". J Soc Psychical Res. 1985

Harribance CC. Sean Harribance: A psychic predicts the future. Port of Spain: Sean Harribance Institute 1994.

Hécaen H, Ajuriaguerra J. 1952. L'Héautoscopie. In: Méconnassiances et hallucinations corporelles. Paris: Masson.

Hécaen H, Green A. 1957. Sur l'héautoscopie. Encephale.

Irwin HJ. 1985. Flight of mind: a psychological study of the out-of- body experience. Metuchen (NJ): Scarecrow Press.

Irwin HJ. An Introduction to Parapsychology. 4th ed. Jefferson, NC: McFarland; 2004.

Irwin HJ, Watt CA. An Introduction to Parapsychology. 5th ed. Jefferson, NC: McFarland; 2007. 22. Green C. Analysis of spontaneous cases. Proc Soc Psychical Res. 1960

Irwin HJ. Extrasensory experiences and the need for Absorption. Parapsychol Rev. 1989

Jensen, C.G. and Parker, A. "Entangled in the womb? A pilot study on the possible physiological connectedness between identical twins with different embryonic backgrounds" EXPLORE 2012

Kittenis MD, Caryl PGC, Stevens P. Distant physiological interaction effects between related and unrelated participants. Paper presented at the 47th Annual Convention of the Parapsychological Association, Vienna, Austria, 2004

Kittenis MD. Event-related EEG correlations between physically isolated participants. (Unpublished doctoral thesis). Scotland: University of Edinburgh; 2007

Krippner S, Persinger M. Evidence for enhanced congruence between dreams and distant target material during periods of decreased geomagnetic activity. J Sci Explor 1996.

Kölmel HW. 1985. Complex visual hallucinations in the hemianopic field. J Neurol Neurosurg Psychiatry.

Lackner JR. 1988. Some proprioceptive influences on the perceptual representation of body shape and orientation. Brain.

Lackner JR. 1992. Sense of body position in parabolic flight. Ann N Y Acad Sci.

Leube DT, Knoblich G, Erb M, Grodd W, Bartels M, Kircher TT. 2003. The neural correlates of perceiving one's own movements. Neuroimage.

Lippman CW. 1953. Hallucination of physical duality in migraine. J Nerv Ment Dis.

Lobel E, Kleine J, Leroy-Wilig A. 1999. Functional MRI of galvanic vestibular stimulation. J Neurophysiol.

Lunn V. 1970. Autoscopic phenomena. Acta Psychiatr Scand 46(Suppl 219).

Melzack R. 1990. Phantom limbs and the concept of a neuromatrix. Trends Neurosci.

Metzinger T. 2003. Being no one. Cambridge (MA): MIT Press.

Mittelstaedt H, Glasauer S. 1993. Illusions of verticality in weightlessness. Clin Invest.

McDonough BE, Don NS, Warren CA. Gamma band ("40 Hz") EEG and unconscious target detection in a psi task. Paper presented at the 43rd Annual Convention of the Parapsychological Association, Freiburg, Germany, 2002

Moulton ST, Kosslyn SM. Using neuroimaging to resolve the psi debate. J Cogn Neurosci. 2008

Martens, P.R. 1994. "Near-Death Experiences in Out-of-Hospital Cardiac Arrest Survivors. Meaningful Pheneomena or just Fantasy of Death?" Resuscitation.

Moody, Paul. 1975. Life after Life: The Investigation of a Phenomenon—Survival of Bodily Death . Atlanta: Mockingbird Books.

Madhurst, R.G. "A telepathy test with eighty-five pairs of identical twins" Journal of the Society of Psychical Research 1968

Moore WD. Three in four Americans believe in paranormal. The Gallup Organization Poll Releases. 2005.` Available at: http://www.gallup.com/poll/16915/Three-Four-Americans-BelieveParanormal.aspx.

Newport F, Strausberg M. Americans' belief in psychic and paranormal phenomena is up over last decade. The Gallup Organization News Service. 2001. Available at: http://www.gallup. com/poll/4483/americans-belief-psychic-paranormal-phenomenaover-last-decade.aspx.

Newman, H.H. "Telepathy between twins" Journal of the Society of Psychical Research 1949

Nash, C.B. and Buzby, D.E. "Extrasensory perception of identical and fraternal twins: comparison of clairvoyance test scores" Journal of Heredity 1965

Newberg, A. "Cerebral blood flow changes associated with different meditation practices and perceived depth of meditation" Psychiatry Research: Neuroimaging 2010.

Neisser U. 1988. The five kinds of self-knowledge. Phil Psychol.

Non-identical monozygotic twins [internet], Develop Biol. Sunderland, MA: Sinauer Associates; 2006. Available at: http://8e.devbio.com/article.php?id111

Naskar, A. The Art of Neuroscience in Everything, 2015

455

Parker A. "A ganzfeld study using identical twins". J Soc Psychical Res. 2010

Playfair GL. Identical twins and telepathy. J Soc Psychical Res. 1999

Prasad J, Stevenson I. A survey of spontaneous psychical experiences in school children of Uttar Pradesh, India. Int J Parapsychol. 1968

Persinger, MA. "Experimental Demonstration of Potential Entanglement of Brain Activity over 300 Km for Pairs of Subjects Sharing the Same Circular Rotating, Angular Accelerating Magnetic Fields: Verification by s_LORETA, QEEG Measurements" Journal of Consciousness Exploration & Research 2013

Playfair GL. Twin Telepathy: the Psychic Connection. London: Vega Books; 2002

Persinger, MA. "Dream ESP Experiments and Geomagnetic Activity" Journal of American Society for Psychical Research Vol 83, 1989.

Persinger, MA. and Saroka, KS. "Protracted parahippocampal activity associated with Sean Harribance" International Journal of Yoga Vol. 5, 2012.

Persinger MA, Krippner S. Dream ESP experiments and geomagnetic activity. J Am Soc Psychical Res 1989.

Persinger MA. Geophysical variables and behaviour: LXXI. Differential contribution of geomagnetic activity to paranormal experiences concerning death and crisis: An alternative to the ESP hypothesis. Percept Motor Skills 1993.

Persinger MA, Saroka KS, Lavallee CF, Booth JN, Hunter MD, Mulligan BP, et al. Correlated cerebral events between physically and sensory isolated pairs of subjects exposed to yoked circumcerebral magnetic fields. Neuroscience Lett 2010.

Persinger MA, Roll WG, Tiller SG, Koren SA, Cook CM. Remote viewing with the artist Ingo Swann: Neuropsychological profile, electroencephalographic correlates, magnetic resonance imaging (MRI) and possible mechanisms. Percept Motor Skills 2002.

Persinger, M. "Billions of Human Brains Immersed Within a Shared Geomagnetic Field: Quantitative Solutions and Implications for Future Adaptations" The Open Biology Journal, 2013.

Persinger MA, Lavallee CF. Theoretical and experimental evidence of macroscopic entanglement between human brain activity and photon emissions; implications for quantum consciousness and future applications. J Cons Explor Res 2010.

Persinger MA, Lavallee CF. The sum of N=N and the quantitative support for the cerebral holographic and electromagnetic configuration of consciousness. J Cons Stud 2012.

Persinger MA. 10-20 Joules as a neuromolecular quantum in medicinal chemistry: an alternative to the myriad molecular pathways? Curr Med Chem 2010.

Persinger MA. On the possible representation of the electromagnetic equivalents of all human memory within the earth's magnetic field: implications for theoretical biology. Theor Biol Insights 2008.

Persinger, MA. "Schumann Resonance Frequencies Found Within Quantitative Electroencephalographic Activity: Implications for Earth-Brain Interactions" International Letters of Chemistry, Physics and Astronomy 2014.

Persinger MA. The harribance effect as pervasive out-of-body experiences. Neuroquantology 2010.

Persinger, MA. "Quantitative Evidence for Direct Effects Between Earth-Ionosphere Schumann Resonances and Human Cerebral Cortical Activity" International Letters of Chemistry, Physics and Astronomy 2014.

Persinger, MA. "Terrestrial and lunar gravitational forces upon the mass of a cell: relevance to cell function" International Letters of Chemistry, Physics and Astronomy 2014.

Palmer J. 1978. The out-of-body experience: a psychological theory. Parapsychol Rev.

Radin DI. Event-related electroencephalographic correlations between isolated human subjects. J Altern Complement Med. 2004

Ruby P, Decety J. 2001. Effect of subjective perspective taking during simulation of action: a PET investigation of agency. Nat Neurosci.

Ryan A. New insights into the links between ESP and geomagnetic activity. J Sci Explor, 2008.

Radin DI. Unconscious perception of future emotions: an experiment in presentiment. J Sci Explor. 1997

Richards TL, Kozak L, Johnson LC, Standish LJ. Replicable functional magnetic resonance imaging evidence of correlated brain signals between physically and sensory isolated subjects. J Altern Complement Med. 2005

Radin DI, Minds E. Extrasensory Experiences in a Quantum Reality. New York: Paraview and Pocket Books; 2006

Roll WG, Williams BJ. Quantum theory, neurobiology, and parapsychology. In: Krippner S, Friedman H. Mysterious Minds: the Neurobiology of Psychics, Mediums and Other Extraordinary People. Santa Barbara: Praeger; 2010

Rhine LE. Hallucinatory psi experiences: I. An introductory survey. J Parapsychol. 1956

Sannwald G. On the psychology of spontaneous paranormal phenomena. Int J Parapsychol. 1963

Schouten SA. Analysing spontaneous cases: a replication based on the Sannwald collection. Eur J Parapsychol. 1981

Schouten SA. Analysing spontaneous cases: a replication based on the Rhine collection. Eur J Parapsychol. 1982

Stevenson I. Telepathic Impressions: A Review and Report of Thirty-Five New Cases. Charlottesville, VA: University Press of Virginia; 1970

Schouten SA. A different approach for analyzing spontaneous cases with particular reference to the study of Louisa E. Rhine's case collection. J Parapsychol. 1983

Sommer, R., Osmond, H. and Pancyr, L. "Selection of twins for ESP experimentation" IJP 1961

Spottiswoode SJP, May EC. Skin conductance prestimulus response: Analyses, artifacts and a pilot study. J Sci Explor. 2003

Shiva Purana Four Volumes, by J.L Shastri edition 1970

Schmidt S, Schneider R, Utts J, Walach H. Distant intentionality and the feeling of being stared at: two meta-analyses. Br J Psychol. 2004

Sabell A, Clarke C, Fenwick P. Inter-subject EEG-correlations at a distance: the transferred potential. Paper Presented at the 44th Annual Convention of the Parapsychological Association, New York, 2001

Sheils D. 1978. A cross-cultural study of beliefs in out-of-the-body experiences, waking and sleeping. J Soc Psych Res.

Smith BH. 1960. Vestibular disturbances in epilepsy. Neurol.

Schumann W. O. (1952). "Über die strahlungslosen Eigenschwingungen einer leitenden Kugel, die von einer Luftschicht und einer Ionosphärenhülle umgeben ist". Zeitschrift und Naturfirschung 7a: 149–154. doi:10.1515/zna-1952-0202.

Schumann W. O. (1952). "Über die Dämpfung der elektromagnetischen Eigenschwingnugen des Systems Erde – Luft – Ionosphäre". Zeitschrift und Naturfirschung 7a: 250–252. doi:10.1515/zna-1952-3-404.

Schumann W. O. (1952). "Über die Ausbreitung sehr Langer elektriseher Wellen um die Signale des Blitzes". Nuovo Cimento 9 (12): 1116–1138. doi:10.1007/BF02782924.

Schumann W. O. & H. König (1954). "Über die Beobactung von Atmospherics bei geringsten Frequenzen". Naturwiss 41 (8): 183–184. doi:10.1007/BF00638174.

Strassman, R. "DMT: The Spirit Molecule" 2001.

Targ E, Schlitz M, Irwin HJ. Psi-related experiences. In: Cardeña E, Lynn SJ, Krippner S (eds). Varieties of Anomalous Experience: Examining the Scientific Evidence. Washington, DC: American Psychological Association; 2004

Venkatasubramanian G, Jayakumar PN, Nagendra HR, Nagaraja D, Deeptha R, Gangadhar BN. Investigating paranormal phenomena: Functional brain imaging of telepathy. Int J Yoga 2008.

Walker EH. The Physics of Consciousness. New York: Basic Books; 2000

Wackermann J, Seiter C, Keibel H, Walach H. Correlations between brain electrical activities of two spatially separated human subjects. Neurosci Lett. 2003

Whinnery, J.E. 1997. "Psychophysiologic Correlates of Unconsciousness and near- death experiences." Journal of Near-Death Studies.

Watson, P "Twins: An Investigation into the Strange Coincidences in the Lives of Separated Twins" 1981

Zacks JM, Ollinger JM, Sheridan MA, Tversky B. 2002. A parametric study of mental spatial transformations of bodies. Neuroimage.

BOOK III
THE GOD PARASITE
REVELATION OF NEUROSCIENCE

Ardila A, Gomez J. Paroxysmal "feeling of somebody being nearby". Epilepsia 1988

Babayev ES, Allahverdiyeva AA. Effects of geomagnetic activity variations on the physiological and psychological state of functionally healthy humans: some results of Azerbaijani studies. Adv Space Res 2007.

Bickerton, D. (2009). Adam's tongue: How humans made language and how language made humans. New York: Hill and Wang.

Brothers, L. (2002). The social brain: A project for integrating primate behavior and neurophysiology in a new domain. In J. T. Cacioppo et al. (Eds.), Foundations in neuroscience. Cambridge, MA: MIT Press.

Bhagavad-gita As It Is, The Bhaktivedanta Book Trust 2010.

Blanke, O. and Arzy, S. "The Out-of-Body Experience: Disturbed Self-Processing at the Temporo-Parietal Junction" THE NEUROSCIENTIST 2005.

Blanke O, Ortigue S, Landis T, Seeck M. 2002. Stimulating illusory own-body perceptions. Nature.

Bonda E, Petrides M, Frey S, Evans A. 1995. Neural correlates of mental transformations of the body-in-space. Proc Natl Acad Sci U S A.

Brandt T. 2000. Central vestibular disorders. In: Vertigo: its multisensory syndromes. London: Springer.

Brugger P, Agosti R, Regard M, Wieser HG, Landis T. 1994. Heautoscopy, epilepsy, and suicide. J Neurol Neurosurg Psychiatr.

Brugger P, Regard M, Landis T. 1997. Illusory reduplication of one's own body: phenomenology and classification of autoscopic phenomena. Cogn Neuropsychiatr.

Burchett, Scott A. and Phillip T. Hicks. 2006. "The mysterious trace amines: Protean neuromodulators of synaptic transmission in mammalian brain." Progress in Neurobiology.

Bancaud, J., Brunet-Bourgin, J., Chauvel, P., & Halgren, E. (1994). Anatomical origin to deja vu and vivid "memories" in human temporal lobe epilepsy. Brain, 117

Belisheva, N. K., Popov, A. N., Petukhova, N. V., Pavlova, L. P., Osipov, K. S., Tkachenko, S.E., & Baranova, T.I. (1995). Quantitative and qualitative evaluations of the effect of geomagnetic variations on the functional state of the brain. Biophysics, 40

Booth, J. N., Koren, S. A., & Persinger, M. A. (2003). Increased proportions of sensed presences and occipital spikes with 1- and 10-msec point durations of continuous 7-Hz transcerebral magnetic fields. Perceptual and Motor Skills, 97

Booth, J. N., Koren, S. A., & Persinger, M. A. (2005). Increased feelings of the sensed presence and increased geomagnetic activity at the time of the experience during transcerebral exposure to weak complex magnetic fields. International Journal of Neuroscience, 115

Cherry, N. (2002). Schumann resonances, a plausible biophysical mechanism for the human health effects of solar/geomagnetic activity. Natural Hazards,26

Cook, C. M., & Persinger, M. A. (1997). Experimental induction of the "sensed presence" in normal subjects and an exceptional subject. Perceptual and Motor Skills, 85

Cook, C. M., & Persinger, M. A. (2001). Geophysical variables and behavior: XCII. Experimental elicitation of the experience of a sentient being by right hemispheric, weak magnetic fields: interaction with temporal lobe sensitivity. Perceptual and Motor Skills, 92

Calvert GA, Campbell R, Brammer MJ. 2000. Evidence from functional magnetic resonance imaging of crossmodal binding in the human heteromodal cortex. Curr Biol.

Classical Hindu Mythology: A Reader in the Sanskrit Puranas, Temple University Press 1978.

Darwin, Charles. "On the origin of species by means of natural selection" (original edition, 1859).

Darwin, Charles. "The Descent of Man" (original edition, 1871).

Daly DD. 1958. Ictal affect. Am J Psychiatry.

Decety J, Sommerville JA. 2003. Shared representations between self and other: a social cognitive neuroscience view. Trends Cogn Sci.

Denning TR, Berrios GE. 1994. Autoscopic phenomena. Br J Psychiatry.

Devinsky O, Feldmann E, Burrowes K, Bromfield E. 1989. Autoscopic phenomena with seizures. Arch Neurol.

Downing PE, Jiang Y, Shuman M, Kanwisher N. 2001. A cortical area selective for visual processing of the human body. Science.

Dewhurst K, Beard AW. Sudden religious conversions in temporal lobe epilepsy. 1970 Epilepsy Behav 2003

Dastur H.M., Desai A.D., "A comparative study of brain tuberculomas and gliomas based upon 107 case records of each". Brain. 1965

Devinsky O, Lai G. Spirituality and religion in epilepsy. Epilepsy Behav 2008.

Dostoyevsky, The Possessed (original edition 1872)

Dostoyevsky, The Brothers Karamazov, (original edition 1880)

Dostoyevsky, The Insulted and Injured, (original edition 1861)

Dostoyevsky, The Idiot,. (original edition 1869)

Edelman, G. M. (1992). Bright air, brilliant fire: On the matter of the mind. New York: Basic Books.

Esquirol, Étienne (1838). Baillière, Jean-Baptiste (and sons), ed. Des maladies mentales considérées sous les rapports médical, hygiénique et médico-légal, [Mental illness as considered in medical, hygienic, and medico-legal reports] Volume 1 and 2.

Esquirol, Étienne 1845. Mental maladies; a treatise on insanity (original French edition 1838).

Farrell MJ, Robertson IH. 2000. The automatic updating of egocentric spatial relationships and its impairment due to right posterior cortical lesions. Neuropsychologia.

Fasold O, von Bevern M, Kuhberg M, Ploner CJ, Vilringer A, Lempert T, Wenzel R. 2002. Human vestibular cortex as identified with caloric vestibular stimulation by functional magnetic resonance imaging. Neuroimage.

Gazzaniga, M. S. (1985). The social brain. New York: Basic Books.

Greenspan, S. I. and S. G. Shanker (2004). The first idea: How symbols, language, and intelligence evolved from our early primate ancestors to modern humans. Cambridge, MA: Da Capo Press.

Goodwin GM, McCloskey DI, Matthews PBC. 1972. Proprioceptive illusions induced by muscle vibration: contribution by muscle spindles to perception? Science.

Good News Bible, New International Version

Green CE. 1968. Out-of-body experiences. London: Hamish Hamilton.

Geschwind N. "Behavioural changes in temporal lobe epilepsy". Psychol Med. 1979.

Greyson, Bruce. 2006. "Near-Death Experiences and Spirituality." Zygon: Journal of Religion and Science.

Grossman E, Donnelly M, Price R, Pickens D, Morgan V, Neighbor G, and others. 2000. Brain areas involved in perception of biological motion. J Cogn Neurosci.

Grüsser OJ, Landis T. 1991. The splitting of "I" and "me": heautoscopy and related phenomena. In: Visual agnosias and other disturbances of visual perception and cognition. Amsterdam: MacMillan.

Guru Granth Sahib -English Version 2012.

Halligan PW. 2002. Phantom limbs: the body in mind. Cogn Neuropsychiatr.

Halligan PW, Fink GR, Marshal JC, Vallar G. 2003. Spatial cognition: evidence from visual neglect. Trends Cogn Sci.

Hécaen H, Ajuriaguerra J. 1952. L'Héautoscopie. In: Méconnassiances et hallucinations corporelles. Paris: Masson.

Hécaen H, Green A. 1957. Sur l'héautoscopie. Encephale.

Harrington A. Unfinished business: models of laterality in the nineteenth century. In: Davidson RJ, Hugdhal K, editors. Brain asymmetry. MIT Press; 1995.

Hobbs, J. (2006). The origins and evolution of language: A plausible strong-AI account. In M. Arbibi (Ed.), Action to language via the mirror neuron system. Cambridge: Cambridge University Press.

Irwin HJ. 1985. Flight of mind: a psychological study of the out-of- body experience. Metuchen (NJ): Scarecrow Press.

Jaspinder Singh, The Sikh Gurus - Lives and Teachings: Spiritual Enlightenment Through Message Of Sikhism, Jawahar Publishers 2014.

Kölmel HW. 1985. Complex visual hallucinations in the hemianopic field. J Neurol Neurosurg Psychiatry.

Lakoff, G. and M. Johnson (1999). Philosophy in the flesh. Basic Books: New York.

LeDoux, J. E. (1996). The emotional brain. New York: Simon & Schuster.

Litovitz TA, Penafiel M,Krause D,Zhang D,Mullins JM. The role of temporal sensing in bioelectromagnetic effects. Bioelectromagnetics 1997.

Lagace N, St-Pierre LS, Persinger MA. Attenuation of epilepsy-induced brain dam- age in the temporal cortices of rats by exposure to LTP-patterned magnetic fields. Neurosci Lett 2009.

Lackner JR. 1988. Some proprioceptive influences on the perceptual representation of body shape and orientation. Brain.

Lackner JR. 1992. Sense of body position in parabolic flight. Ann N Y Acad Sci.

Leube DT, Knoblich G, Erb M, Grodd W, Bartels M, Kircher TT. 2003. The neural correlates of perceiving one's own movements. Neuroimage.

Lippman CW. 1953. Hallucination of physical duality in migraine. J Nerv Ment Dis.

Lobel E, Kleine J, Leroy-Wilig A. 1999. Functional MRI of galvanic vestibular stimulation. J Neurophysiol.

Lunn V. 1970. Autoscopic phenomena. Acta Psychiatr Scand 46

Lama Surya Das, Awakening the Buddha Within: Tibetan Wisdom for the Western World, Broadway Books; Reprint edition 1998.

Makarec, K., & Persinger, M. A. (1985). Temporal lobe signs: Electroencephalographic validity and enhanced scores in special populations. Perceptual and Motor Skills, 60

Makarec, K., & Persinger, M. A. (1990). Electroencephalographic validation of a temporal lobe signs inventory in a normal population. Journal of Research in Personality, 24

Melzack R. 1990. Phantom limbs and the concept of a neuromatrix. Trends Neurosci.

Metzinger T. 2003. Being no one. Cambridge (MA): MIT Press.

Mittelstaedt H, Glasauer S. 1993. Illusions of verticality in weightlessness. Clin Invest.

Maryansky, A. (1996). African Ape social structure: A blue print for reconstructing early hominid structure. In J. Steel, S. Sherman (Eds.), The Archeology of Human Ancestry. London: Rutledge.

Massey, D. (2000). What I don't know about my field but wish I did. Annual Review of Sociology.

Massey, D. S. (2002). A brief history of human society: The origin and role of emotion in social life: 2001 presidential address. American Sociological Review.

Miller, B. D. (2007). Cultural anthropology, 4th ed. Boston: Allyn & Bacon.

Moody, Paul. 1975. Life after Life: The Investigation of a Phenomenon—Survival of Bodily Death. Atlanta: Mockingbird Books.

Mulligan BP, Hunter MD, Persinger MA. Effects of geomagnetic activity and atmospheric power variations on quantitative measures of brain activity: replication of the Azerbaijani studies. Adv Space Res 2010.

Martens, P.R. 1994. "Near-Death Experiences in Out-of-Hospital Cardiac Arrest Survivors. Meaningful Pheneomena or just Fantasy of Death?" Resuscitation.

Novembre, J., J. K. Pritchard and G. Coop (2007). Adaptive drool in the gene pool. Nature Genetics.

Newberg, A. "Cerebral blood flow changes associated with different meditation practices and perceived depth of meditation" Psychiatry Research: Neuroimaging 2010.

Neisser U. 1988. The five kinds of self-knowledge. Phil Psychol.

Pepperberg, I. (2008). Alex and me. HarperCollins: New York.

Persinger, "'I would kill in God's name' role of sex, weekly church attendance, report of a religious experience and limbic lability" Perceptual and Motor Skills 1997.

Persinger "Experimental simulation of the God experience" Neurotheology 2003.

Persinger, M. A. (1993b). Personality changes following brain injury as a grief response to the loss of sense of self: Phenomenological themes as indices of local lability and neurocognitive restructuring as psycho- therapy. Psychological Reports, 72

Persinger, Corradini, Clement, Keaney, et al "Neurotheology and its convergence with neuroquantology" NeuroQuantology 2010.

Persinger, Koren and St-Pierre "The electromagnetic induction of mystical and altered states within the laboratory" Journal of Consciousness Exploration and Research 2010.

Persinger "Case report: A prototypical spontaneous 'sensed presence' of a sentient being and concomitant electroencephalographic activity in the clinical laboratory" Neurocase 2008.

Persinger and Saroka "Potential production of Hughlings Jackson's "parasitic consciousness" by physiologically-patterned weak transcerebral magnetic fields: QEEG and source localization" Epilepsy & Behavior 28 (2013).

Persinger. "The neuropsychiatry of paranormal experiences". J Neuropsychiatry Clin Neurosci 2001.

Persinger. "Neuropsychological bases of god beliefs", New York: Praeger, 1987

Persinger. "Temporal lobe epileptic signs and correlative behaviors displayed by normal populations", Journal of General Psychology, 1986

Persinger "Experimental Facilitation of the Sensed Presence: Possible Intercalation between the Hemispheres Induced by Complex Magnetic Fields" Journal of Nervous and Mental Disease 2002.

Palmer J. 1978. The out-of-body experience: a psychological theory. Parapsychol Rev.

Richardson, K. (1999). The making of intelligence. London: Phoenix.

Ratnasuriya, R.H. "Joan of Arc, creative psychopath: is there another explanation?" Journal of The Royal Society of Medicine 1986.

Rilling, J. K. (2006). Human and nonhuman primate brains: Are they allometrically scaled versions of the same design? Evolutionary Anthropology.

Ruby P, Decety J. 2001. Effect of subjective perspective taking during simulation of action: a PET investigation of agency. Nat Neurosci.

Roth, G and Dicke, U,(2005) Evolution of the Brain and the Intelligence, TRENDS in Cognitive Sciences.

Sawer, G. and Deak, V. (2007). The last human. New York: Peter N. Nevraumont Publication – Yale University Press.

Small, D. (2008). On the deep history of the brain. Berkeley: University of California Press.

Sheils D. 1978. A cross-cultural study of beliefs in out-of-the-body experiences, waking and sleeping. J Soc Psych Res.

Smith BH. 1960. Vestibular disturbances in epilepsy. Neurol.

Srimad-Bhagavatam, The Bhaktivedanta Book Trust 2012.

Strassman, R. "DMT: The Spirit Molecule" 2001.

Slater E, Beard AW. The schizophrenia-like psychoses of epilepsy. Br J Psychiatry 1963.

Turner, B. (2000a). Embodied ethnography. Doing culture. Social Anthropology.

Turner, J. H. (2000b). On the origins of human emotions: A sociological inquiry into the evolution of human affect. Stanford, California: Stanford University Press.

The Quran, Translated by Muhammad Abdel Haleem, Oxford University Press 2008.

The Upanishads (Classic of Indian Spirituality), Nilgiri Press 2009.

The Upanishads (Penguin Classics), Penguin Classics; Reissue edition 1965.

The Bhagavad Gita (Classics of Indian Spirituality), Nilgiri Press 2007.

The Mahabharata (Penguin Classics), Penguin Classics; Abridged edition 2009.

The Book of Mormon: Another Testament of Jesus Christ, Church of Jesus Christ of Latter Day Saints 1981.

Tiller, S. G., & Persinger, M. A. (2002). Geophysical variables and behavior: XCVII. Increased proportions of left-sided sense of presence induced experimentally by right hemispheric application of specific (frequency-modulated) complex magnetic fields. Perceptual and Motor Skills, 94

Tononi, G., & Edelman, G. E. (1998). Consciousness and complexity. Science, 282

Whinnery, J.E. 1997. "Psychophysiologic Correlates of Unconsciousness and near- death experiences." Journal of Near-Death Studies.

Zacks JM, Ollinger JM, Sheridan MA, Tversky B. 2002. A parametric study of mental spatial transformations of bodies. Neuroimage.

Book IV
The Spirituality Engine

Ardila A, Gomez J. Paroxysmal "feeling of somebody being nearby". Epilepsia 1988.

Agarwal, S. Genocide of Women in Hinduism, Sudrastan Books 1999.

Alexander King and Bertrand Schneider, The first global revolution New York: Simon and Schuster, 1991.

Anthropos, "Das Problem des Totemismus", Volumes 9,10,11.

Bickerton, D. (2009). Adam's tongue: How humans made language and how language made humans. New York: Hill and Wang.

Brothers, L. (2002). The social brain: A project for integrating primate behavior and neurophysiology in a new domain. In J. T. Cacioppo et al. (Eds.), Foundations in neuroscience. Cambridge, MA: MIT Press.

Bhagavad-gita As It Is, The Bhaktivedanta Book Trust 2010.

Blanke, O. and Arzy, S. "The Out-of-Body Experience: Disturbed Self-Processing at the Temporo-Parietal Junction" THE NEUROSCIENTIST 2005.

Blanke O, Ortigue S, Landis T, Seeck M. 2002. Stimulating illusory own-body perceptions. Nature.

Bird-David, Nurit. "'Animism' Revisited: Personhood, Environment, and Relational Epistemology." Current Anthropology 40 Supplement (1999).

Boas, F. "The Origin of Totemism", American Anthropologist, Volume 18, 1916

Bonda E, Petrides M, Frey S, Evans A. 1995. Neural correlates of mental transformations of the body-in-space. Proc Natl Acad Sci U S A.

Brandt T. 2000. Central vestibular disorders. In: Vertigo: its multisensory syndromes. London: Springer.

Brugger P, Agosti R, Regard M, Wieser HG, Landis T. 1994. Heautoscopy, epilepsy, and suicide. J Neurol Neurosurg Psychiatr.

Brugger P, Regard M, Landis T. 1997. Illusory reduplication of one's own body: phenomenology and classification of autoscopic phe- nomena. Cogn Neuropsychiatr.

Calvert GA, Campbell R, Brammer MJ. 2000. Evidence from functional magnetic resonance imaging of crossmodal binding in the human heteromodal cortex. Curr Biol.

Classical Hindu Mythology: A Reader in the Sanskrit Puranas, Temple University Press 1978.

Callaway, Henry. The Religious System of the Amazulu. Springvale: Springvale Mission, 1868–1870; Cape Town: Struik, 1970.

Dawkins, R. "The God Delusion" 2006

Dawkins, R. "The Selfish Gene 1976"

Darwin, Charles. "On the origin of species by means of natural selection" (original edition, 1859).

Darwin, Charles. "The Descent of Man" (original edition, 1871).

Daly DD. 1958. Ictal affect. Am J Psychiatry.

Decety J, Sommerville JA. 2003. Shared representations between self and other: a social cognitive neuroscience view. Trends Cogn Sci.

Denning TR, Berrios GE. 1994. Autoscopic phenomena. Br J Psychiatry.

Devinsky O, Feldmann E, Burrowes K, Bromfield E. 1989. Autoscopic phenomena with seizures. Arch Neurol.

Downing PE, Jiang Y, Shuman M, Kanwisher N. 2001. A cortical area selective for visual processing of the human body. Science.

Dastur H.M., Desai A.D., "A comparative study of brain tuberculomas and gliomas based upon 107 case records of each". Brain. 1965

Devinsky O, Lai G. Spirituality and religion in epilepsy. Epilepsy Behav 2008.

Edelman, G. M. (1992). Bright air, brilliant fire: On the matter of the mind. New York: Basic Books.

Frazer, J. G., Totemism and Exogamy. 4 vols. London, 1910.

Freud, S. (1905) "Three essays on the Theory of Sexuality"

Freud, S. (1924) "Totem and Taboo"

Freud, S. (1927) "Fetishism"

"Form and Content in Totemism," American Anthropologist, Vol. 20, 1918.

Farrell MJ, Robertson IH. 2000. The automatic updating of egocentric spatial relationships and its impairment due to right posterior cor- tical lesions. Neuropsychologia.

Fasold O, von Bevern M, Kuhberg M, Ploner CJ, Vilringer A, Lempert T, Wenzel R. 2002. Human vestibular cortex as identified with caloric vestibular stimulation by functional magnetic resonance imaging. Neuroimage.

Gazzaniga, M. S. (1985). The social brain. New York: Basic Books.

Greenspan, S. I. and S. G. Shanker (2004). The first idea: How symbols, language, and intelligence evolved from our early primate ancestors to modern humans. Cambridge, MA: Da Capo Press.

Goodwin GM, McCloskey DI, Matthews PBC. 1972. Proprioceptive illusions induced by muscle vibration: contribution by muscle spindles to perception? Science.

Good News Bible, New International Version

Green CE. 1968. Out-of-body experiences. London: Hamish Hamilton.

Geschwind N. "Behavioural changes in temporal lobe epilepsy". Psychol Med. 1979.

Greyson, Bruce. 2006. "Near-Death Experiences and Spirituality." Zygon: Journal of Religion and Science.

Grossman E, Donnelly M, Price R, Pickens D, Morgan V, Neighbor G, and others. 2000. Brain areas involved in perception of biological motion. J Cogn Neurosci.

Grüsser OJ, Landis T. 1991. The splitting of "I" and "me": heautoscopy and related phenomena. In: Visual agnosias and other disturbances of visual perception and cognition. Amsterdam: MacMillan.

Guthrie, Stewart Elliot. Faces in the Clouds: A New Theory of Religion. New York: Oxford University Press, 1993.

Goldenweiser, A., "Totemism, an Analytical Study," Journal of American Folklore, Vol. XXIII, 1910

Handbook of American Indians North of Mexico. Bureau of American Ethnology, Smithsonian Institution, Bulletin 30, 2 vols. Washington, 1907-1910.

Hilger, M. I., "Some Early Customs of the Menomini Indians," Journal de la Societe des Americanistes, Vol. XLIX 1960.

Halligan PW. 2002. Phantom limbs: the body in mind. Cogn Neuropsychiatr.

Halligan PW, Fink GR, Marshal JC, Vallar G. 2003. Spatial cognition: evidence from visual neglect. Trends Cogn Sci.

Hécaen H, Ajuriaguerra J. 1952. L'Héautoscopie. In: Méconnassiances et hallucinations corporelles. Paris: Masson.

Hobbs, J. (2006). The origins and evolution of language: A plausible strong-AI account. In M. Arbibi (Ed.), Action to language via the mirror neuron system. Cambridge: Cambridge University Press.

Irwin HJ. 1985. Flight of mind: a psychological study of the out-of- body experience. Metuchen (NJ): Scarecrow Press.

Ibn Warraq, What the Koran Really Says, Prometheus Books 2002

Jakobson, R. and Halle, M., Fundamentals of Language. 's-Gravenhage, 1956.

Jenness, D., "The Ojibwa Indians of Parry Island: Their Social and Religious Life," Bulletin of the Canadian Department of Mines, Ottawa, 1935.

Kölmel HW. 1985. Complex visual hallucinations in the hemianopic field. J Neurol Neurosurg Psychiatry.

Kinietz, W. V., "Chippewa Village: The Story of Katikitegon," Bulletin of the Cranbrook Institute of Science, Detroit, 1947.

Kroeber, A. L., "Totem and Taboo: An Ethnologic Psychoanalysis," (1920) reprinted in The Nature of Culture, Chicago, 1952. --,Anthropology. New York, 1923.

Landes, R., "Ojibwa Sociology," Columbia University Contributions to Anthropology, Vol. XXIX, New York, 1937.

Lane, B. S., "Varieties of Cross-cousin Marriage and Incest Taboos: Structure and Causality," Essays in the Science of Culture, ed. G. E. Dole and R. L. Carneiro, New York, 1960.

Levi-Strauss, C., Les Structures elementaires de la parente. Paris, 1949. --,La Pensee Sauvage, Paris, 1962.

Linton, R., "Totemism and the A. E. F.," American Anthropologist, Vol. 26, 1924.

Long, J. K, Voyages and Travels of an Indian Interpreter and Trader [1791], Chicago, 1922.

Lubbock, John. The Origin of Civilization and the Primitive Condition of Man. London: Longmans, Green, 1889 (orig. edn, 1870).

Lowie, R. H., "On the Principle of Convergence in Ethnology," Journal of American Folklore, Vol. XXV, 1912. --, Primitive Society. Reprinted 1947. New York, 1920. --,An Introduction to Cultural Anthropology.

Lakoff, G. and M. Johnson (1999). Philosophy in the flesh. Basic Books: New York.

LeDoux, J. E. (1996). The emotional brain. New York: Simon & Schuster.

Litovitz TA, Penafiel M,Krause D,Zhang D,Mullins JM.The role of temporal sensing in bioelectromagnetic effects. Bioelectromagnetics 1997.

Lagace N, St-Pierre LS, Persinger MA. Attenuation of epilepsy-induced brain dam- age in the temporal cortices of rats by exposure to LTP-patterned magnetic fields. Neurosci Lett 2009.

Lackner JR. 1988. Some proprioceptive influences on the perceptual representation of body shape and orientation. Brain.

Lackner JR. 1992. Sense of body position in parabolic flight. Ann N Y Acad Sci.

Leube DT, Knoblich G, Erb M, Grodd W, Bartels M, Kircher TT. 2003. The neural correlates of perceiving one's own movements. Neuroimage.

Lippman CW. 1953. Hallucination of physical duality in migraine. J Nerv Ment Dis.

Lobel E, Kleine J, Leroy-Wilig A. 1999. Functional MRI of galvanic vestibular stimulation. J Neurophysiol.

Lunn V. 1970. Autoscopic phenomena. Acta Psychiatr Scand 46(Suppl 219).

Metzinger T. 2003. Being no one. Cambridge (MA): MIT Press.

Mittelstaedt H, Glasauer S. 1993. Illusions of verticality in weightlessness. Clin Invest.

Maryansky, A. (1996). African Ape social structure: A blue print for reconstructing early hominid structure. In J. Steel, S. Sherman (Eds.), The Archeology of Human Ancestry. London: Rutledge.

Massey, D. (2000). What I don't know about my field but wish I did. Annual Review of Sociology.

Massey, D. S. (2002). A brief history of human society: The origin and role of emotion in social life: 2001 presidential address. American Sociological Review.

Miller, B. D. (2007). Cultural anthropology, 4th ed. Boston: Allyn & Bacon.

Moody, Paul. 1975. Life after Life: The Investigation of a Phenomenon—Survival of Bodily Death . Atlanta: Mockingbird Books.

Mulligan BP, Hunter MD, Persinger MA. Effects of geomagnetic activity and atmospheric power variations on quantitative measures of brain activity: replication of the Azerbaijani studies. Adv Space Res 2010.

Martens, P.R. 1994. "Near-Death Experiences in Out-of-Hospital Cardiac Arrest Survivors. Meaningful Pheneomena or just Fantasy of Death?" Resuscitation.

Marx, Karl. Capital, Volume 1 (original edition, 1867).

Masuzawa, Tomoko. "Troubles with Materiality: The Ghost of Fetishism in the Nineteenth Century." Comparative Studies in Society and History 42 (2000)

M'Lennan, John Ferguson. "The Worship of Animals and Plants." Fortnightly Review 6 (1868); 7 (1870)

McConnel, U., "The Wik-Munkan tribe of Cape York Peninsula," Oceania, Vol. I, 1930.

McLennan, J. F., "The Worship of Animals and Plants," Fortnightly Review, Vols. 6 and 7, 1869-1870.

Malan, V. D. and McCone, R. C., "The Time Concept Perspective and Premise in the Socio-cultural Order of the Dakota Indians," Plains Anthropologist, Vol. 5, 1960.

Malinowski, B., The Sexual Life of Savages in North-western Melanesia. 2 vols. New York-London, 1929. --, Magic, Science and Religion. Boston, 1948.

Michelson, T., "Explorations and Fieldwork of the Smithsonian Institution in 1925," Smithsonian Miscellaneous Collections, Vol. 78, No. I. Washington, 1926.

Murdock, G. P., Social Structure. New York, 1949. Notes and Queries on Anthropology. Sixth edition. London, 1951.

Naskar, A. "The God Parasite: Revelation of Neuroscience" 2015

Naskar, A. "The Art of Neuroscience in Everything" 2015

Novembre, J., J. K. Pritchard and G. Coop (2007). Adaptive drool in the gene pool. Nature Genetics.

Newberg, A. "Cerebral blood flow changes associated with different meditation practices and perceived depth of meditation" Psychiatry Research: Neuroimaging 2010.

Neisser U. 1988. The five kinds of self-knowledge. Phil Psychol.

New York, 1934. --, Social Organization. New York, 1948.

Principe, W. "Towards defining spirituality", Studies in religion/Sciences religieuses 12/2 (1983)

Pietz, William. "The Problem of the Fetish, I." Res: Anthropology and Aesthetics 9 (1985).

Piddington, R., An Introduction to Social Anthropology, Vol, I. Edinburgh-London.

Prytz Johansen, J., The Maori and His Religion and Its Non-ritualistic Aspects. Copenhagen, 1954.

Pepperberg, I. (2008). Alex and me. HarperCollins: New York.

Persinger, "'I would kill in God's name' role of sex, weekly church attendance, report of a religious experience and limbic lability" Perceptual and Motor Skills 1997.

Persinger "Experimental simulation of the God experience" Neurotheology 2003.

Persinger, Corradini, Clement, Keaney, et al "Neurotheology and its convergence with neuroquantology" NeuroQuantology 2010.

Persinger, Koren and St-Pierre "The electromagnetic induction of mystical and altered states within the laboratory" Journal of Consciousness Exploration and Research 2010.

Persinger "Case report: A prototypical spontaneous 'sensed presence' of a sentient being and concomitant electroencephalographic activity in the clinical laboratory" Neurocase 2008.

Persinger and Saroka "Potential production of Hughlings Jackson's "parasitic consciousness" by physiologically-patterned weak transcerebral magnetic fields: QEEG and source localization" Epilepsy & Behavior 28 (2013).

Persinger. "The neuropsychiatry of paranormal experiences". J Neuropsychiatry Clin Neurosci 2001.

Persinger "Experimental Facilitation of the Sensed Presence: Possible Intercalation between the Hemispheres Induced by Complex Magnetic Fields" Journal of Nervous and Mental Disease 2002.

Palmer J. 1978. The out-of-body experience: a psychological theory. Parapsychol Rev.

Richardson, K. (1999). The making of intelligence. London: Phoenix.

Ratnasuriya, R.H. "Joan of Arc, creative psychopath: is there another explanation?" Journal of The Royal Society of Medicine 1986.

Rafiqul-Haqq M. and Newton P. Women in Islam.

Radcliffe-Brown, A. R., "The Sociological Theory of Totemllm111 (1929) reprinted in Structure and Function in Primitive Sool1t) London, 1952.

Rilling, J. K. (2006). Human and nonhuman primate brains: Are they allometrically scaled versions of the same design? Evolutionary Anthropology.

Ruby P, Decety J. 2001. Effect of subjective perspective taking during simulation of action: a PET investigation of agency. Nat Neurosci.

Roth, G and Dicke, U,(2005) Evolution of the Brain and the Intelligence, TRENDS in Cognitive Sciences.

Sawer, G. and Deak, V. (2007). The last human. New York: Peter N. Nevraumont Publication – Yale University Press.

Small, D. (2008). On the deep history of the brain. Berkeley: University of California Press.

Sheils D. 1978. A cross-cultural study of beliefs in out-of-the-body experiences, waking and sleeping. J Soc Psych Res.

Smith BH. 1960. Vestibular disturbances in epilepsy. Neurol.

Srimad-Bhagavatam, The Bhaktivedanta Book Trust 2012.

Strassman, R. "DMT: The Spirit Molecule" 2001.

Slater E, Beard AW. The schizophrenia-like psychoses of epilepsy. Br J Psychiatry 1963.

Stanner, W. E. H., "Murinbata kinship and totemism," Oceania, Vol. 7. 1936-1937.

Sarkar, P.R. The Awakening of Women, Ananda Marga Publciations, 1995.

Tylor, E.B. "The Limits of Savage Religion." Journal of the Royal Anthropological Institute 21 (1892)

Tylor, E.B. Primitive Culture, 2 vols. London: John Murray, 1871.

"Totem and Taboo in Retrospect," (1939) reprinted in The Nature of Culture, Chicago, 1952. --,Anthropology. New edition. New York, 1948.

"Taboo," (1939) reprinted in Structure and Function in Primitive Society. London, 1952.

"The Comparative Method in Social Anthropology," Journal of the Royal Anthropological Institute, Vol. 81, 1951, reprinted as Chapter V in Method in Social Anthropology. Chicago, 1958.

Thomas, N. W., Kinship Organizations and Group Marriage in Australia. Cambridge, 1906.

Tylor, E. B., "Remarks on Totemism with Especial Reference to some Modem Theories Concerning It," Journal of the Royal Anthropological Institute, Vol. XXVIII, 1899.

The Rig Veda, Classic Century Works 2012

The Holy Vedas ; Rig Veda, Yajur Veda, Sama Veda, Atharva Veda, B.R. Publishing Corporation 2006

The Law Code of Manu (Oxford World's Classics), Oxford University Press; Reissue edition 2009

Turner, B. (2000a). Embodied ethnography. Doing culture. Social Anthropology.

Turner, J. H. (2000b). On the origins of human emotions: A sociological inquiry into the evolution of human affect. Stanford, California: Stanford University Press.

The Quran, Translated by Muhammad Abdel Haleem, Oxford University Press 2008.

The Upanishads (Classic of Indian Spirituality), Nilgiri Press 2009.

The Upanishads (Penguin Classics), Penguin Classics; Reissue edition 1965.

The Bhagavad Gita (Classics of Indian Spirituality), Nilgiri Press 2007.

The Mahabharata (Penguin Classics), Penguin Classics; Abridged edition 2009.

The Book of Mormon: Another Testament of Jesus Christ, Church of Jesus Christ of Latter Day Saints 1981.

Van Gennep, A., L'Etat actuel du probleme totemique. Paris, 1920.

Van Rheenan, Gailyn. Communicating Christ in Animistic Contexts. Grand Rapids, MI: Baker Book House, 1991.

Warren, W., "History of the Ojibways," Collections of the Minnesota Historical Society, Vol. V. Saint Paul, Minn., 1885.

Werner, W. L., A Black Civilization. Revised edition. New York, 1958.

Whinnery, J.E. 1997. "Psychophysiologic Correlates of Unconsciousness and near- death experiences." Journal of Near-Death Studies.

Zelenine, D., Le Culte des idoles en Siberie. Paris, 1952.

Zacks JM, Ollinger JM, Sheridan MA, Tversky B. 2002. A parametric study of mental spatial transformations of bodies. Neuroimage.

BOOK V
LOVE SUTRA
THE NEUROSCIENTIFIC MANUAL OF LOVE

Acevedo BP, Aron A, Fisher HE & Brown LL (2011). Neural correlates of long-term intense romantic love. Social Cognitive and Affective Neuroscience, published online January 5 2011 doi:10.1093/scan/nsq092.

Aron A (2006). Relationship neuroscience: Advancing the social psychology of close relationships using functional neuroimaging. In PAM Van Lange (Ed) Bridging social psychology: Benefits of transdisciplinary approaches, Lawrence Erlbaum Associates Publishers Mahwah NJ.

Aron A (2010). Behavior the brain and the social psychology of close relationships. In CR Agnew, DE Carlston, WG Graziano & JR Kelly (Eds) Then a miracle occurs: Focusing on behavior in social psychological theory and research, Oxford University Press New York.

Aron A, Fisher H, Mashek DJ, Strong G, Li H & Brown LL (2005). Reward motivation and emotion systems associated with early-stage intense romantic love. Journal of Neurophysiology 94.

Aron AP & Aron EN (1986). Love as the expansion of self: Understanding attraction and satisfaction. Hemisphere, New York.

Aron AP & Aron EN (1991). Love and sexuality. In K McKinney & S Sprecher (Eds.) Sexuality in close relationship. Lawrence Erlbaum Associates, Hillsdale, NJ.

Bartels A & Zeki S (2000). The neural basis of romantic love. Neuroreport: For Rapid Communication of Neuroscience Research 11.

Bartels A & Zeki S (2004). The neural correlates of maternal and romantic love. Neuroimage 21.

Basson R (2000). The female sexual response: A different model. Journal of Sex & Marital Therapy 26, 51-65.

Basson R (2002). Women's sexual desire: Disordered or misunderstood? Journal of Sex & Marital Therapy 28.

Basson R, Wierman ME, van Lankveld J & Brotto L (2010). Summary of the recommendations on sexual dysfunctions in women. Journal of Sexual Medicine 7.

Baumeister RF (2000). Gender differences in erotic plasticity: The female sex drive as socially flexible and responsive. Psychological Bulletin 126.

Beauregard M, Courtemanche J, Paquette V & St-Pierre EL (2009). The neural basis of unconditional love. Psychiatry Research: Neuroimaging 172.

Bianchi-Demicheli F, Grafton ST & Ortigue S (2006). The power of love on the human brain. Social Neuroscience 1.

Bocher M, Chisin R, Parag Y, Freedman N, Meir Weil Y, Lester H et al. (2001). Cerebral activation associated with sexual arousal in response to a pornographic clip: A 15O-H2O PET study in heterosexual men. Neuroimage 14.

Brotto LA, Bitzer J, Laan E, Leiblum SR & Luria M (2010). Women's sexual desire and arousal disorders. Journal of Sexual Medicine 7.

Carter CS (1998). Neuroendocrine perspectives on social attachment and love. Psychoneuroendocrinology 23.

Carter CS & Keverne EB (2002). The neurobiology of social affiliation and pair bonding. In J Pfaff AP Arnold AE Etgen & SE Fahrbach (Eds) Hormones brain and behavior, vol. 1. Academic Press, New York.

Campbell A. The limbic system and emotion in relation to acupuncture. Acupuncture in Medicine.1999.

Cowan CP, Cowan PA. When Partners Become Parents: The Big Life Change for Couples. New York: Basic Books; 1992.

Chivers ML & Bailey JM (2005). A sex difference in features that elicit genital response. Biological Psychology 70.

Chivers ML, Rieger G, Latty E & Bailey JM (2004). A sex difference in the specificity of sexual arousal. Psychological Science 15.

Chivers ML Seto MC & Blanchard R (2007). Gender and sexual orientation differences in sexual response to sexual activities versus gender of actors in sexual films. Journal of Personality and Social Psychology 93.

Cook, R. "Mirror neurons: From origin to function" Behavioral and Brain Sciences 2014.

Chatterton RT, Jr, Hill PD, Aldag JC, Hodges KR, Belknap SM, Zinaman MJ. Relation of plasma oxytocin and prolactin concentrations to milk production in mothers of preterm infants: Influence of stress. J Clin Endocrinol Metab. 2000.

Dixson, Barnaby J. and Brookes, Robert C. "The role of facial hair in women's perceptions of men's attractiveness, health, masculinity and parenting abilities" Evolution and Human Behavior 34, 2013

Diamond LM (2003). What does sexual orientation orient? A biobehavioral model distinguishing romantic love and sexual desire. Psychological Review 110.

Diamond LM (2005). From the heart or the gut? Sexual-minority women's experiences of desire for same-sex and other-sex partners. Feminism and Psychology 15

Diamond LM (2008). Sexual fluidity: Understanding women's love and desire. Harvard University Press, Cambridge, MA.

Diamond LM & Wallen K (2011). Sexual-minority women's sexual motivation around the time of ovulation. Archives of Sexual Behavior 40

Esch, T. and Stefano, G.B "The Neurobiology of Love" Neuroendocrinology Letters No.3 June Vol.26, 2005.

Esch T. [Health in stress: Change in the stress concept and its significance for prevention, health and life style]. Gesundheitswesen 2002.

Ferretti A, Caulo M, Del Gratta C, Di Matteo R, Merla A, Montorsi F et al. (2005). Dynamics of male sexual arousal: distinct components of brain activation revealed by fMRI. Neuroimage 26

Fisher HE (1998). Lust attraction and attachment in mammalian reproduction. Human Nature 9

Fonteille V & Stoleru S. (2010). The cerebral correlates of sexual desire: Functional neuroimaging approach. Sexologies 10.1016/j.sexol.2010.03.011.

Francis DD, Champagne FC, Meaney MJ. Variations in maternal behavior are associated with differences in oxytocin receptor levels in the rat. J Neuroendocrinol. 2000;12. [PubMed]

Feldman R. Infant-mother and infant-father synchrony: The co-regulation of positive arousal. Infant Ment Health J. 2003;24.

Feldman R. Parent-infant synchrony and the construction of shared timing; physiological precursors, developmental outcomes, and risk conditions. J Child Psychol Psychiatry. 2007;48.

Gordon I, Zagoory-Sharon O, Leckman JF, Feldman R. "Oxytocin and the development of parenting in humans." Biol Psychiatry. 2010 Aug 15;68(4). doi: 10.1016/j.biopsych.2010.02.005.

Glezerman, M. (2006). Five years to the term breech trial: The rise and fall of a randomized controlled trial. Am J Obstet Gynecol, 194(1)

Garstein, M. (2003). "Studying infant temperament." Infant Behavior and Development

Gatewood, J. D., and M. D. Morgan, et al. (2005). "Motherhood mitigates aging- related decrements in learning and memory and positively affects brain aging in the rat." Brain Res Bull 66 (2)

Genazzani, A. D. (2005). "Neuroendocrine aspects of amenorrhea related to stress." Pediatr Endocrinol Rev 2 (4)

Getchell, T. (1991). Smell and Taste in Health and Disease. New York: Raven Press.

Giammanco, M., G. Tabacchi, et al. (2005). "Testosterone and aggressiveness." Med Sci Monit 11 (4)

Giedd, J. (2005). Personal communication. Giedd, J. N. (2003). "The anatomy of mentalization: A view from developmental neuroimaging." Bull Menninger Clin 67

Haheim, L. L., Albrechtsen, S., Berge, L. N., Bordahl, P. E., Egeland, T., Henriksen, T., et al. (2004). Breech birth at term: Vaginal delivery or elective cesarean section? A systematic review of the literature by a norwegian review team. Acta Obstet Gynecol Scand, 83

Hannah, M. E., Whyte, H., Hannah, W. J., Hewson, S., Amankwah, K., Cheng, M., et al. (2004). Maternal outcomes at 2 years after planned cesarean section versus planned vaginal birth for breech presentation at term: The international randomized term breech trial. Am J Obstet Gynecol, 191

Hemanth P. Nair and Larry J. Young "Vasopressin and Pair-Bond Formation: Genes to Brain to Behavior" Physiology Published 1 April 2006 Vol. 21 no. 2, DOI: 10.1152/physiol.00049.2005

Haith MM, Bergman T, Moore MJ. Eye contact and face scanning in early infancy. Science. 1977.

Hrdy, S. B. (1997). "Raising Darwin's consciousness: Female sexuality and the prehominid origins of patriarchy." Human Nature 8

Hrdy, S. B. (2000). "The optimal number of fathers: Evolution, demography, and history in the shaping of female mate preferences." Ann NY Acad Sci 907

Huber, D., P. Veinante, et al. (2005). "Vasopressin and oxytocin excite distinct neuronal populations in the central amygdala." Science 308

Hultcrantz, M. (2006). "Estrogen and hearing: A summary of recent investigations." Acta Otolaryngol 126

Hamann S Herman RA Nolan CL & Wallen K (2004). Men and women differ in amygdala response to visual sexual stimuli. Nature Neuroscience 7.

Hatfield E & Sprecher S. (1986). Measuring passionate love in intimate relationships. Journal of Adolescence 9

Hatfield E. Love, Sex and Intimacy. New York: Harper Collins 1993.

Heaton JP, Adams MA. Update on central function relevant to sex: remodeling the basis of drug treatments for sex and the brain. Int J Impot Res 2003.

Heinzel A, Walter M, Schneider F, Rotte M, Matthiae C, Tempelmann C et al. (2006). Self-related processing in the sexual domain: Parametric event-related fMRI study reveals neural activity in ventral cortical midline structures. Social Neuroscience 1

Insel TR and Hulihan TJ. A gender-specific mechanism for pair bonding: oxytocin and partner preference formation in monogamous voles. Behav Neurosci 109, 1995.CrossRefMedlineWeb of Science

Neumann ID. Brain oxytocin: A key regulator of emotional and social behaviours in both females and males. J Neuroendocrinol. 2008.

Insel TR and Shapiro LE. Oxytocin receptor distribution reflects social organization in monogamous and polygamous voles. Proc Natl Acad Sci USA 89, 1992.

Insel TR, Wang ZX, and Ferris CF. Patterns of brain vasopressin receptor distribution associated with social organization in microtine rodents. J Neurosci 14, 1994.

Jeong GW, Park K, Youn G, Kang HK, Kim HJ, Seo JJ et al. (2005). Assessment of cerebrocortical regions associated with sexual arousal in premenopausal and menopausal women by using BOLD-based functional MRI. The Journal of Sexual Medicine 2.

Kierse MJNC. (2002) Evidence-based childbirth only for breech babies? Birth, 29(1).

Kotaska, A. (2004). Inappropriate use of randomised trials to evaluate complex phenomena: Case study of vaginal breech delivery. BMJ, 329(7473).

Karama S, Lecours AR, Leroux J-M, Bourgouin P, Beaudoin G, Joubert S et al. (2002). Areas of brain activation in males and females during viewing of erotic film excerpts. Human Brain Mapping 16.

Kelley AE (2004). Ventral striatal control of appetitive motivation: Role in ingestive behavior and reward-related learning. Neuroscience and Biobehavioral Reviews 27.

Komisaruk BR, Whipple B. Love as sensory stimulation: physicological consequences of its deprivation and expression. Psychoneuroendocrinology 1998.

Lacoboni, M. and Dapretto, M. "The mirror neuron system and the consequences of its dysfunction" Nature 2005.

Lamb ME. A re-examination of the infant social world. Hum Dev. 1977;20.

Maravilla KR & Yang CC (2007). Sex and the brain: The role of fMRI for assessment of sexual function and response. International Journal of Impotence Research 19.

Maravilla KR & Yang CC (2008). Magnetic resonance imaging and the female sexual response: Overview of techniques results and future directions. Journal of Sexual Medicine 5.

Masters WH & Johnson VE (1966). Human sexual response. Boston: Little Brown.

Maternal, Infant and Reproductive Health Research Unit (MIRU). Choosing Delivery by Caesarean: Has Its Time Come? Conference sponsored by MIRU, University of Toronto, Toronto, Ontario, Canada, Nov 7, 2002.

Moulier V, Mouras H, Pelegrini-Isaac M, Glutron D, Rouxel R & Grandjean B (2006). Neuroanatomical correlates of penile erection evoked by photographic stimuli in human males. Neuroimage 33.

Naskar, A. "The Art of Neuroscience in Everything" 2015

Naskar, A. "Introduction to Ellynizer - An Advanced Quantum Biological Device for Eliminating Menstrual Problems" Global Journal of Medical Research Vol 14, No 5 2014

O'Doherty JP (2004). Reward representations and reward-related learning in the human brain: Insights from neuroimaging. Current Opinion in Neurobiology 14.

Ortigue S & Bianchi-Demicheli F (2007). Interactions between human sexual arousal and sexual desire: a challenge for social neuroscience. Revue Medicale Suisse 3.

Ortigue S, Bianchi-Demicheli F de C, Hamilton AF & Grafton ST (2007). The neural basis of love as a subliminal prime: An event-related functional magnetic resonance imaging study. Journal of Cognitive Neuroscience 19.

Ortigue S, Bianchi-Demicheli F, Patel N, Frum C & Lewis JW (2010). Neuroimaging of love: fMRI meta-analysis evidence toward new perspectives in sexual medicine. Journal of Sexual Medicine 7.

Ortigue S, Patel N & Bianchi-Demicheli F (2009). New electroencephalogram (EEG) neuroimaging methods of analyzing brain activity applicable to the study of human sexual response. Journal of Sexual Medicine 6.

Onions CT. The Oxford Dictionary of English Etymology. New York: Oxford University Press 1966.

Phan KL, Wager T, Taylor SF & Liberzon I (2002). Functional neuroanatomy of emotion: A meta-analysis of emotion activation studies in PET and fMRI. Neuroimage 16.

Redoute J, Stoleru S, Gregoire MC, Costes N, Cinotti L, Lavenne F et al. (2000). Brain processing of visual sexual stimuli in human males. Human Brain Mapping 11.

Redoute Jr M, Stolery S, Gregoire M-C, Costes N, Cinotti L, Lavenne F et al. (2000). Brain processing of visual sexual stimuli in human males. Human Brain Mapping 11.

Regan PC (1998). Of lust and love: Beliefs about the role of sexual desire in romantic relationships. Personal Relationships 5.

Rieger G, Bailey JM & Chivers ML (2005). Sexual arousal patterns of bisexual men. Psychological Science 16.

Robson KS, Moss HA. Patterns and determinants of maternal attachment. J Pediatr. 1970.

"RCOG statement on umbilical non-severance or 'lotus birth'". Royal College of Obstetricians and Gynaecologists. 1 December 2008.

Rust PCR (2000). Bisexuality in the United States: A reader and guide to the literature. Columbia University Press, New York.

Rizzolatti, G. & Fadiga, L. (1998) Grasping objects and grasping action meanings: The dual role of monkey rostroventral premotor cortex (area F5). Novartis Foundation Symposium 218.

Rizzolatti, G., Fadiga, L., Gallese, V. & Fogassi, L. (1996) Premotor cortex and the recognition of motor actions. Social Cognitive and Affective Neuroscience 3 (2).

Rizzolatti, G., Fogassi, L. & Gallese, V. (2001) Neurophysiological mechanisms underlying the understanding and imitation of action. Nature Reviews Neuroscience 2(9).

Rizzolatti G., Fogassi L. & Gallese V. (2004) Cortical mechanism subserving object grasping, action understanding and imitation. In: The cognitive neurosciences, 3rd edition, ed. M. S. Gazzaniga, A Bradford Book/MIT Press

Rizzolatti, G. & Arbib, M. A. (1998) Language within our grasp. Trends in Neurosciences 21(5). [aRC, LLH]

Rizzolatti, G., Camarda, R., Fogassi, L., Gentilucci, M., Luppino, G. & Matelli, M. (1988) Functional organization of inferior area 6 in the macaque monkey. II. Area F5 and the control of distal movements. Experimental Brain Research 71 (3).

Rizzolatti, G. & Luppino, G. (2001) The cortical motor system. Neuron 31.

Rizzolatti, G. & Matelli, M. (2003) Two different streams form the dorsal visual system: Anatomy and functions. Experimental Brain Research 153.

Rizzolatti, G. & Sinigaglia, C. (2008) Mirrors in the brain. How our minds share actions and emotions. Oxford University Press.

Rizzolatti, G. & Sinigaglia, C. (2010) The functional role of the parieto-frontal mirror circuit: Interpretations and misinterpretations. Nature Reviews Neuroscience 11 (4).

Ramachandran, V. S. (2000) Mirror neurons and imitation learning as the driving force behind "the great leap forward" in human evolution. Edge69. [Available Online at: http://www.edge.org/3rd_culture/ramachandran/rama chandran_in- dex.html]

Ramachandran,V. S. (2009) The neurons that shaped civilization. Available at: http:// www.ted.com/talks/vs_ramachandran_the_neurons_th at_shaped_civilization.Html

Rocca, M. A., Tortorella, P., Ceccarelli, A., Falini, A., Tango, D., Scotti, G., Comi, G. & Fillipi, M. (2008) The "mirror-neuron system" in MS: A 3 tesla fMRI study. Neurology 70(4).

Rochat, M. J., Caruana, F., Jezzini, A., Escola, L., Intskirveli, I., Grammont, F., Gallese, V., Rizzolatti, G. & Umiltà, M. A. (2010) Responses of mirror neurons in area F5 to hand and tool grasping observation. Experimental Brain Research 204(4).

Rochat, M. J., Serra, E., Fadiga, L. & Gallese, V. (2008) The evolution of social cognition: Goal familiarity shapes monkeys' action understanding. Current Biology 18(3).

Rochat, P. (1998) Self-perception and action in infancy. Experimental Brain Research 123.

Rosenbaum, D. (1991) Human motor control. Academic Press.

Roth, T. L. (2012) Epigenetics of neurobiology and behavior during development and adulthood. Developmental Psychobiology 54(6). doi: 10.1002/dev.20550.

Rushworth, M. F., Mars, R. B. & Sallet, J. (2013) Are there specialized circuits for social cognition and are they unique to humans? Current Opinion in Neurobiology 23(3).

Russell, J. L., Lyn, H., Schaeffer, J. A. & Hopkins, W. D. (2011) The role of socio- communicative rearing environments in the development of social and physical cognition in apes. Developmental Science 14(6).

Stern DN. The First Relationship: Infant and Mother. Cambridge, MA: Harvard University Press; 1977.

Sampson, G. (2002) Exploring the richness of the stimulus. The Linguistic Review 19.

Stefano GB, Scharrer B. Endogenous morphine and related opiates, a new class of chemical messengers. Adv Neuroimmunol 1994

Stefano GB, Scharrer B, Smith EM, Hughes TK, Magazine HI, Bil- finger TV et al. Opioid and opiate immunoregulatory processes. Crit Rev in Immunol 1996.

Simpson JA, Rholes WS. Stress and secure base relationships in adulthood. In: Bartholomew K, Perlman D, editors. Advances in personal relationships (Vol. 5): Attachment processes in adult- hood. London: Kingsley 1994

Shipley JT. Dictionary of Word Origins. New York: Philosophical Library 1945.

Slingsby BT, Stefano GB. Placebo: Harnessing the power within. Modern Aspects of Immunobiology 2000

Slingsby BT, Stefano GB. The active ingredients in the sugar pill: Trust and belief. Placebo 2001

Small DM, Jones-Gotman M, Dagher A. Feeding-induced dopamine release in dorsal striatum correlates with meal pleas- antness ratings in healthy human volunteers. Neuroimage 2003

Small DM, Zatorre RJ, Dagher A, Evans AC, Jones-Gotman M. Changes in brain activity related to eating chocolate: From pleasure to aversion. Brain 2001

Smith CM. Elements of Molecular Neurobiology. 3rd ed. New York: Wiley-Liss 2002

Sonetti D, Peruzzi E, Stefano GB. Endogenous morphine and ACTH association in neural tissues. Medical Science Monitor 2005

Spector S, Munjal I, Schmidt DE. Endogenous morphine and codeine. Possible role as endogenous anticonvulsants. Brain Res 2001

Spencer H. Principles of Psychology. New York: Appleton 1800

Stefano GB. Endocannabinoid immune and vascular signaling. Acta Pharmacologica Sinica 2000

Stefano GB, Benson H, Fricchione GL, Esch T. The Stress Re- sponse: Always good and when it is bad. New York: Medical Science International 2005

Stefano GB, Cadet P, Zhu W, Rialas CM, Mantione K, Benz D et al. The blueprint for stress can be found in invertebrates. Neuroendocrinology Letters 2002.

Sanefuji, W. & Ohgami, H. (2013) "Being-imitated" strategy at home-based intervention for young children with autism. Infant Mental Health Journal 34(1). doi: 10.1002/imhj.21375.

Singer T, Critchley HD & Preuschoff K (2009). A common role of insula in feelings empathy and uncertainty. Trends in Cognitive Sciences 13.

Stoleru S & Mouras H (2007). Brain functional imaging studies of sexual desire and arousal in human males. In E. Janssen (Ed) The psychophysiology of sex. Indiana University Press, Bloomington, IN.

Su, M., Hannah, W. J., Willan, A., Ross, S., & Hannah, M. E. (2004). Planned caesarean section decreases the risk of adverse perinatal outcome due to both labour and delivery complications in the term breech trial. BJOG, 111(10).

Thompson EM & Morgan EM (2008). "Mostly straight" young women: Variations in sexual behavior and identity development. Developmental Psychology 44.

Uvnas-Moberg K, Petersson M. [Oxytocin, a mediator of antistress, well-being, social interaction, growth and healing] Z Psychosom Med Psychother. 2005;51.

Van Roosmalen, J., & Rosendaal, F. (2002). There is still room for disagreement about vaginal delivery of breech infants at term. BJOG, 109.

Whyte, H., Hannah, M. E., Saigal, S., Hannah, W. J., Hewson, S., Amankwah, K., et al. (2004). Outcomes of children at 2 years after planned cesarean birth versus planned vaginal birth for breech presentation at term: The international randomized term breech trial. Am J Obstet Gynecol, 191(3).

Walter M, Bermpohl F, Mouras H, Schiltz K, Tempelmann C, Rotte M et al. (2008). Distinguishing specific sexual and general emotional effects in fMRI-subcortical and cortical arousal during erotic picture viewing. Neuroimage 40.

Weinberg MS, Williams CJ & Pryor DW (1994). Dual attraction: Understanding bisexuality. Oxford University Press, New York.

Xu X, Aron A, Brown L, Cao G, Feng T & Weng X (2010). Reward and motivation systems: A brain mapping study of early-stage intense romantic love in Chinese participants. Human Brain Mapping 32.

Young, Larry J. "The Neural Basis of Pair Bonding in a Monogamous Species: A Model for Understanding the Biological Basis of Human Behavior" Offspring: Human Fertility Behavior in Biodemographic Perspective. National Academies Press (US); 2003.

Young, Larry J. and Wang, Z. "The neurobiology of pair bonding" Nature Neuroscience 7 (2004)